Grade **6**

KUMON MATH WORKBOOKS

Geometry & Measurement

Table of Contents

KUM☉N

1 Sort the following numbers into even and odd numbers.   10 points for completion

6, 9, 11, 18, 25, 36, 41, 50, 63, 74, 87

Odd ( )

Even ( )

2 Write the appropriate numbers below.   5 points per question

(1) The number you get from adding 5, 2 tenths, 9 hundredths, and 3 thousandths.

( )

(2) The number you get from adding 4 tenths, 2 hundredths, and 7 thousandths.

( )

3 Write the following fractions in descending order.   5 points per question

(1) $\frac{4}{7}$, $\frac{2}{7}$, $\frac{6}{7}$, $\frac{5}{7}$ ( )

(2) $\frac{3}{8}$, $\frac{3}{5}$, $\frac{3}{4}$, $\frac{3}{10}$ ( )

4 Rewrite each division problem below as a fraction.   5 points per question

(1) $4 \div 7 =$ ( )   (2) $6 \div 5 =$ ( )

(3) $6 \div 9 =$ ( )   (4) $8 \div 6 =$ ( )

5 Line C intersects the parallel lines A and B.   5 points per question

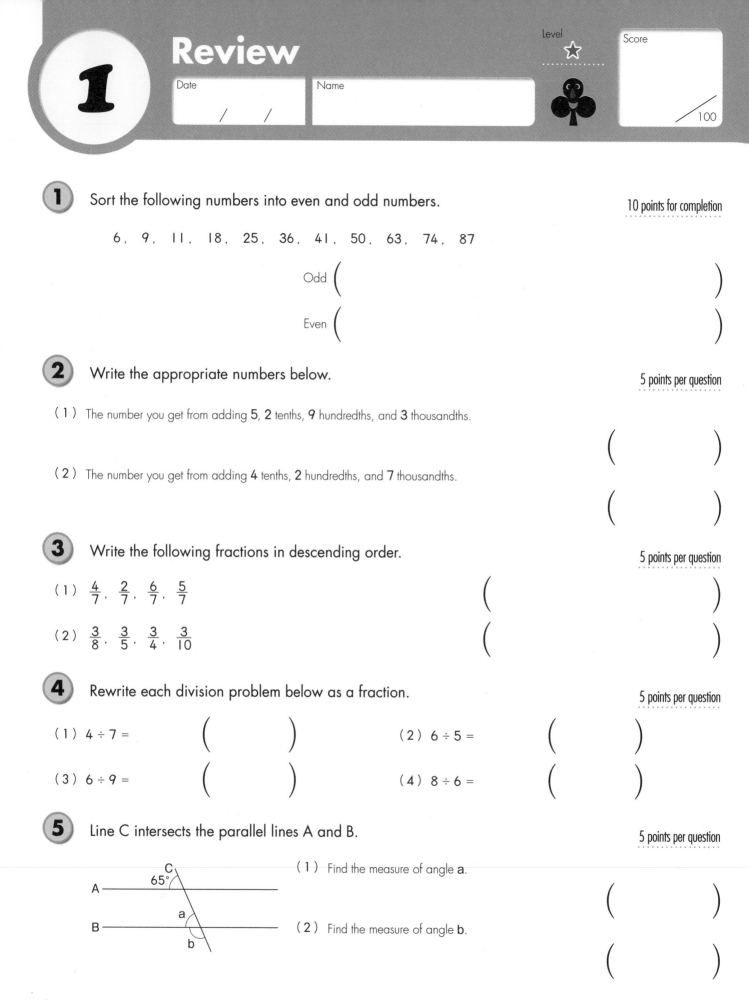

(1) Find the measure of angle **a**.

( )

(2) Find the measure of angle **b**.

( )

6 The quadrilateral below is a parallelogram.

5 points per question

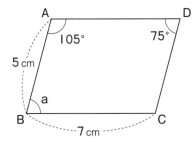

(1) Find the measure of angle **a**.

(   )

(2) How long is side **AD**?

(   )

 7 Find the measure of each angle a below.

5 points per question

(1)

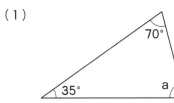

(   )

(2)

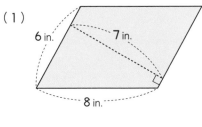

(   )

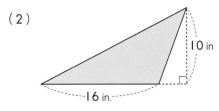

8 Find the area of each shape below.

10 points per question

(1)

(   )

(2)

(   )

Do you remember this? Good!

3

Review

2

Level ☆

Score

/100

Date / /

Name

1 You have three cards each with either a 1, 2 or 3 on them. Use the cards to answer the questions below.

5 points per question

(1) What is the smallest even number you can make with the cards?

( )

(2) What is the largest odd number you can make with the cards?

( )

2 Write the appropriate numbers below.

4 points per question

(1) 10 times 3.4   ( )

(2) 100 times 0.67   ( )

(3) One tenth of 45   ( )

(4) One hundredth of 82.9   ( )

3 Convert each decimal into a fraction and each fraction into a decimal.

4 points per question

(1) $\frac{4}{5}$ =

(2) $\frac{3}{8}$ =

(3) 0.6 =

(4) 1.08 =

4 Circle the larger number.

4 points per question

(1) [0.5 $\frac{3}{5}$]

(2) [1.27 $\frac{5}{4}$]

5 Find the measure of angle a in the figures below.

5 points per question

(1)

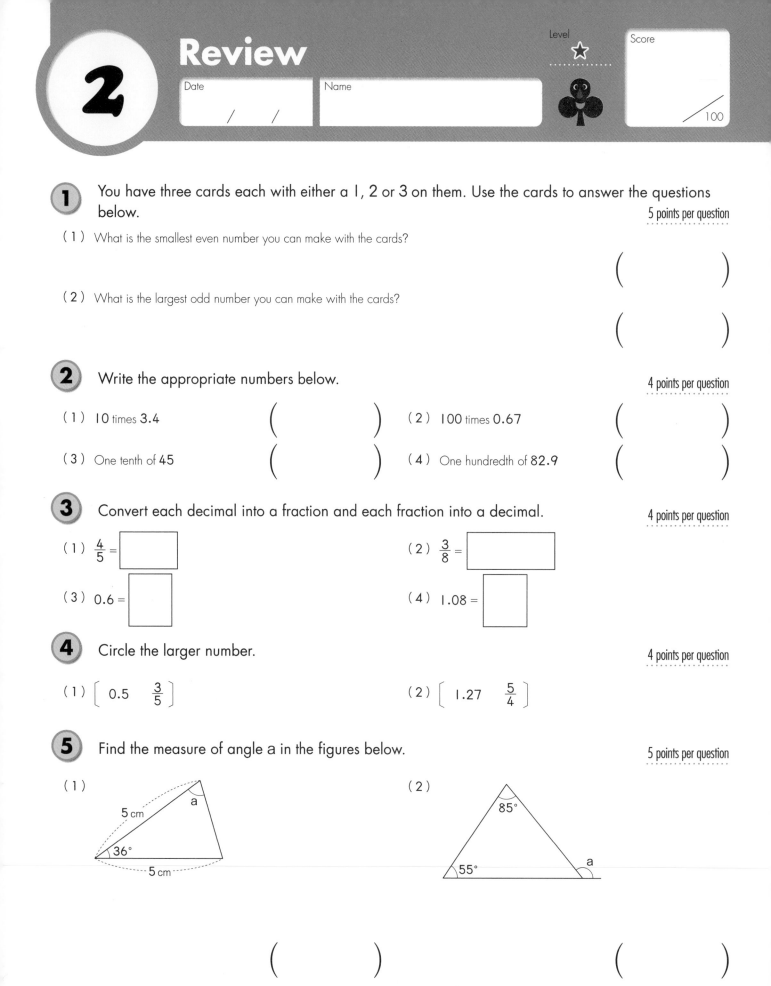

5 cm, a, 36°, 5 cm

(2)

85°, 55°, a

( )

( )

6 Use the letter next to each quadrilateral to answer the questions below.

5 points per question

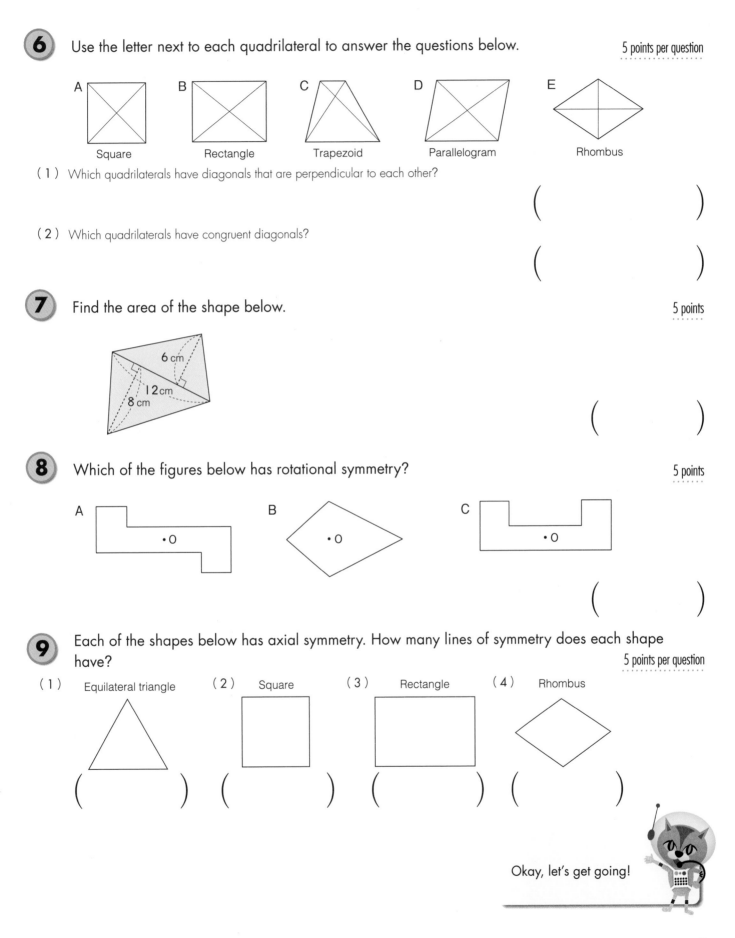

A   B   C   D   E

Square   Rectangle   Trapezoid   Parallelogram   Rhombus

(1) Which quadrilaterals have diagonals that are perpendicular to each other?

(   )

(2) Which quadrilaterals have congruent diagonals?

(   )

7 Find the area of the shape below.

5 points

6 cm

12 cm
8 cm

(   )

8 Which of the figures below has rotational symmetry?

5 points

A   • O   B   • O   C   • O

(   )

9 Each of the shapes below has axial symmetry. How many lines of symmetry does each shape have?

5 points per question

(1) Equilateral triangle   (2) Square   (3) Rectangle   (4) Rhombus

(   )   (   )   (   )   (   )

Okay, let's get going!

3 Least Common Multiple

Level ☆☆

Date / /

Name

Score /100

> **Don't forget!**
>
> The numbers 3, 6 and 9 are **multiples** of 3.

1 Write the multiples of 3 in ascending order.

5 points for completion

$$3 \rightarrow 6 \rightarrow \boxed{} \rightarrow \boxed{} \rightarrow \boxed{} \rightarrow \boxed{}$$

2 Write the multiples of each number below in ascending order.

5 points per question

(1) Multiples of 2 $\quad 2 \rightarrow \boxed{} \rightarrow \boxed{} \rightarrow \boxed{} \rightarrow \boxed{}$

(2) Multiples of 4 $\quad 4 \rightarrow \boxed{} \rightarrow \boxed{} \rightarrow \boxed{} \rightarrow \boxed{}$

(3) Multiples of 5 $\quad 5 \rightarrow \boxed{} \rightarrow \boxed{} \rightarrow \boxed{} \rightarrow \boxed{}$

(4) Multiples of 6 $\quad 6 \rightarrow \boxed{} \rightarrow \boxed{} \rightarrow \boxed{} \rightarrow \boxed{}$

3 Write the multiples of 7 between 1 and 30.

5 points for completion

(   )

4 Write the multiples of 8 between 1 and 30.

5 points for completion

(   )

5 Write the multiples of 9 between 1 and 30.

5 points for completion

(   )

6 Circle the multiples of 3 in the list of numbers below.

5 points for completion

16, 17, 18, 19, 20, 21, 22, 23, 24, 25, 26, 27

┌─ **Don't forget!** ─────────────────────────────────
│ Multiples of 3 can all be divided evenly by 3.
└──

7 Circle the multiples of 3 in the list of numbers below.

5 points for completion

34, 15, 48, 29, 11, 36, 28, 41, 51, 12, 44, 66

8 Circle the multiples of 4 in the list of numbers below.

5 points for completion

24, 32, 35, 36, 38, 40, 42, 44, 46, 48, 50, 52

9 Circle the multiples of 5 in the list of numbers below.

5 points for completion

32, 35, 38, 40, 42, 45, 48, 50, 52, 55, 60

10 How many multiples of 3 are there between 1 and 30?

10 points

(   )

11 How many multiples of 3 are there between 1 and 50?

10 points

(   )

12 How many multiples of 3 are there between 1 and 100?

10 points

(   )

13 How many multiples of 4 are there between 1 and 50?

10 points

(   )

Let's get the ball rolling!

Least Common Multiple

Level ☆☆

Date   /   /

Name

Score   /100

1 Write the multiples of 4 in ascending order below.

5 points for completion

4, 8, 12, ☐, ☐, ☐, ☐, ☐, ☐ ☐

2 Write the multiples of 6 in ascending order below.

5 points for completion

6, 12, ☐, ☐, ☐, ☐, ☐, ☐, ☐

3 Write the multiples of 9 in ascending order below.

5 points for completion

9, ☐, ☐, ☐, ☐, ☐

4 Write the numbers that are multiples of both 4 and 6 in ascending order below.

5 points for completion

☐, ☐, ☐, ☐, ☐

Don't forget!

The numbers that are multiples of both 4 and 6 are called the **common multiples** of 4 and 6.

5 Write the common multiples of 6 and 9 in ascending order below.

5 points for completion

☐, ☐, ☐, ☐

6 Write the common multiples of 4 and 8 in ascending order below.

5 points for completion

☐, ☐, ☐, ☐

© Kumon Publishing Co., Ltd.

7 Write the appropriate number in each box below.

6 points per question

(1) The smallest number in the list of common multiples of **6** and **9** is ☐ .

(2) The smallest number in the list of common multiples of **4** and **8** is ☐ .

Don't forget!

The smallest multiple that two or more numbers have in common is called the **least common multiple (LCM)**.

8 Find the least common multiple of each pair of numbers below.

5 points per question

(1) (6, 8) → ☐

(2) (3, 4) → ☐

(3) (3, 6) → ☐

(4) (4, 6) → ☐

9 Write the appropriate numbers below in order to find the least common multiple of 3, 6 and 9.

6 points per question

(1) The least common multiple of **3** and **6** is ( ) .

The least common multiple of (6) and **9** is ☐ .

(2) The least common multiple of **6** and **9** is ( ) .

The least common multiple of (18) and **3** is ☐ .

(3) The least common multiple of **3** and **9** is ( ) .

The least common multiple of ( ) and **6** is ☐ .

10 Find the least common multiple of each group below.

5 points per question

(1) (2, 3, 8) → ☐

(2) (4, 6, 10) → ☐

(3) (4, 6, 9) → ☐

(4) (6, 8, 12) → ☐

Not so bad, right?

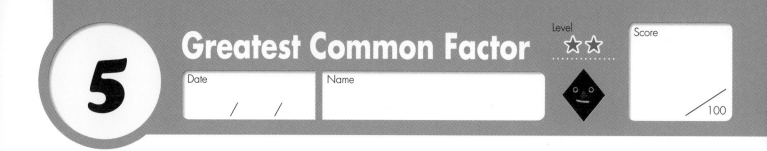

Don't forget!

The number 12 is divisible by the numbers 1, 2, 3, 4, 6 and 12. These numbers are called **factors** of 12.

1 Write the factors of each number below in ascending order.

3 points per question

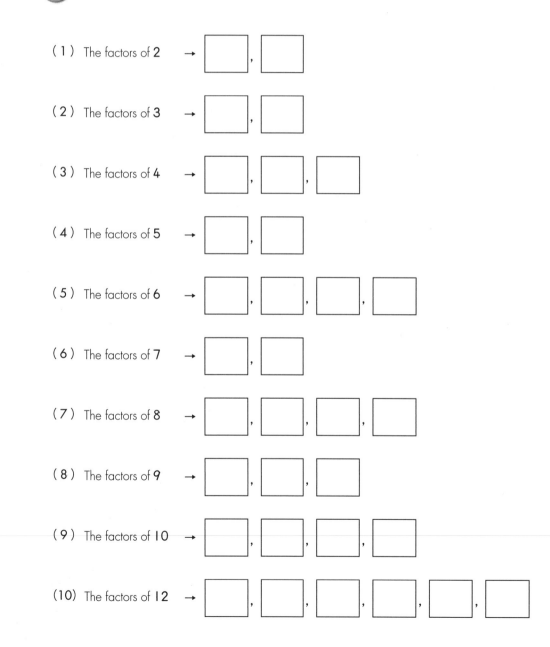

(1) The factors of 2   →   ☐ , ☐

(2) The factors of 3   →   ☐ , ☐

(3) The factors of 4   →   ☐ , ☐ , ☐

(4) The factors of 5   →   ☐ , ☐

(5) The factors of 6   →   ☐ , ☐ , ☐ , ☐

(6) The factors of 7   →   ☐ , ☐

(7) The factors of 8   →   ☐ , ☐ , ☐ , ☐

(8) The factors of 9   →   ☐ , ☐ , ☐

(9) The factors of 10 →   ☐ , ☐ , ☐ , ☐

(10) The factors of 12 →   ☐ , ☐ , ☐ , ☐ , ☐ , ☐

2 Write the appropriate number in each box below.

5 points per question

(1) The factors of 12 are 1, 2, 3, 4, ☐, 12.

(2) The factors of 18 are 1, 2, 3, ☐, 9, 18.

(3) The factors that 12 and 18 have in common are 1, 2, 3, ☐.

3 Write the appropriate number in each box below.

5 points per question

(1) The factors of 20 are 1, 2, 4, ☐, 10, 20.

(2) The factors of 30 are 1, 2, 3, 5, ☐, 10, 15, 30.

(3) The factors that 20 and 30 have in common are 1, 2, 5, ☐.

Don't forget!

The factors that two numbers have in common are called **common factors**.

4 Write the appropriate number in each box below.

5 points per question

(1) The largest number in the list of common factors of 12 and 18 is ☐.

(2) The largest number in the list of common factors of 20 and 30 is ☐.

Don't forget!

The largest factor that two or more numbers have in common is called the **greatest common factor (GCF)**.

5 Find the greatest common factor for each pair of numbers below.

5 points per question

(1) (8, 12) → ☐   (2) (18, 30) → ☐

(3) (12, 30) → ☐   (4) (24, 40) → ☐

(5) (30, 50) → ☐   (6) (24, 64) → ☐

Got your GCF and LCM down? Good!

1 Find the least common multiple for each pair of numbers below.

(1) (2, 3) → ☐   (2) (3, 5) → ☐

(3) (4, 6) → ☐   (4) (3, 9) → ☐

(5) (8, 12) → ☐   (6) (9, 18) → ☐

(7) (12, 18) → ☐   (8) (15, 45) → ☐

(9) (18, 72) → ☐   (10) (16, 80) → ☐

2 Find the least common multiple for each group of numbers below.

(1) (2, 3, 4) → ☐   (2) (3, 4, 5) → ☐

(3) (3, 6, 9) → ☐   (4) (4, 8, 12) → ☐

(5) (5, 10, 15) → ☐   (6) (8, 12, 24) → ☐

Don't forget!

Here's a way to find the least common multiple (LCM).

⟨Example⟩   (12, 18) → ☐ 36

(First, divide by common factors.)

```
2 )12  18  ← Divide by 2.
3 ) 6   9  ← Divide by 3.
    2   3  ← You cannot divide any futher.
```

$2 \times 3 \times 2 \times 3 =$ ☐ 36

(Then, multiply the resulting numbers.)

3 Find the greatest common factor for each pair of numbers below.

(1) (2, 4) → ☐

(2) (4, 8) → ☐

(3) (6, 9) → ☐

(4) (6, 12) → ☐

(5) (9, 15) → ☐

(6) (14, 21) → ☐

(7) (18, 27) → ☐

(8) (20, 25) → ☐

(9) (12, 30) → ☐

(10) (12, 18) → ☐

(11) (16, 24) → ☐

(12) (14, 35) → ☐

(13) (10, 15) → ☐

(14) (30, 32) → ☐

(15) (10, 20) → ☐

(16) (12, 48) → ☐

(17) (45, 60) → ☐

(18) (40, 60) → ☐

(19) (36, 54) → ☐

(20) (60, 120) → ☐

Okay, let's move on then!

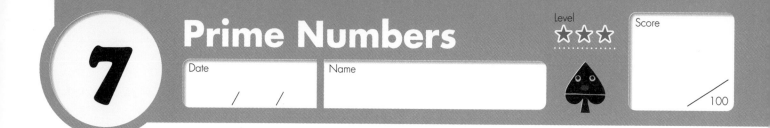

Prime Numbers

Level
☆☆☆

Score

Date
/ /

Name

/100

1 Answer the questions about the numbers 1 through 40 in the number table pictured below.

4 points per question

1	2	3	4	5	6	7	8	9	10
11	12	13	14	15	16	17	18	19	20
21	22	23	24	25	26	27	28	29	30
31	32	33	34	35	36	37	38	39	40

(1) Circle the number **2** and then cross out the multiples of **2**.

(2) Circle the number **3**, and then cross out the multiples of **3**.

(3) Circle the number **5**, and then cross out the multiples of **5**.

(4) Circle the number **7**, and then cross out the multiples of **7**.

(5) Write all the numbers that are circled or not crossed out from the number line above except 1.

(   )

Don't forget!

A **prime number** is a number that has exactly two different factors: 1 and itself. The numbers 2, 3, 5 and 7 are examples of prime numbers.

---

2 Answer the questions below about the numbers 41 and 42.

4 points per question

(1) List all of the factors of **41**.

(   )

(2) Is **41** a prime number?

(   )

(3) List all of the factors of **42**.

(   )

(4) Is **42** a prime number?

(   )

3 Write all of the prime numbers from the list of numbers below.   4 points for completion

43, 45, 47, 51, 54, 57, 59, 61, 65, 67, 69

(   )

4 Write the appropriate number in the box for each number sentence below.   4 points per question

(1) 6 = 2 × ☐

(2) 15 = 3 × ☐

(3) 4 = ☐ × 2

(4) 9 = ☐ × 3

(5) 12 = 2 × 2 × ☐

(6) 8 = 2 × 2 × ☐

Don't forget!

Any integer can be represented as the product of prime numbers.

---

5 Use only prime number factors, write a number sentence for each number below.   3 points per question

(1) 10 =

(2) 12 =

(3) 18 =

(4) 20 =

(5) 21 =

(6) 25 =

(7) 27 =

(8) 28 =

(9) 45 =

(10) 52 =

(11) 60 =

(12) 66 =

Wow! Now we're going to put all of that together and work with fractions.

Fincolumns

Date / /

Name

---

1 The area of each square below is 1 square inch. How many square inches are in each shaded section?

5 points per question

(1)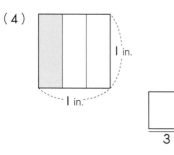

□/2 in.²

(2)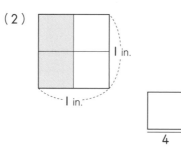

□/4 in.²

(3)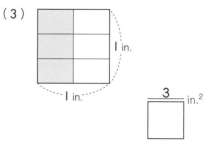

3/□ in.²

(4)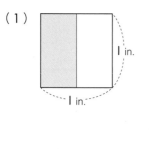

□/3 in.²

(5)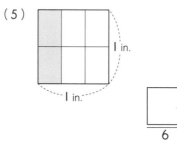

□/6 in.²

(6)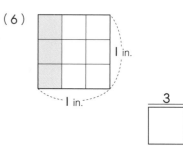

3/□ in.²

(7)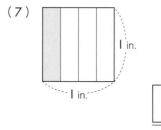

□/4 in.²

(8)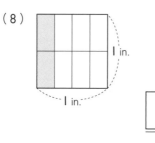

□/8 in.²

(9)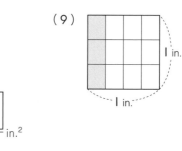

3/□ in.²

---

2 Using the images above as a guide, answer the questions about the fractions below.

3 points per question

(1) $\frac{1}{3}$ = $\frac{\square}{6}$

(2) $\frac{1}{4}$ = $\frac{\square}{8}$

(3) $\frac{2}{4}$ = $\frac{\square}{2}$

(4) $\frac{2}{8}$ = $\frac{\square}{4}$

(5) $\frac{2}{6}$ = $\frac{\square}{3}$

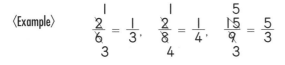

Don't forget!

If you divide the numerator and the denominator by the same number, the result is equal to the first fraction. Using this strategy to find the simplest form of a fraction is called **reduction**.

⟨Example⟩　$\dfrac{\overset{1}{\cancel{2}}}{\underset{3}{\cancel{6}}} = \dfrac{1}{3}$,　$\dfrac{\overset{1}{\cancel{2}}}{\underset{4}{\cancel{8}}} = \dfrac{1}{4}$,　$\dfrac{\overset{5}{\cancel{15}}}{\underset{3}{\cancel{9}}} = \dfrac{5}{3}$

3 Write the appropriate number in each box below.

2 points per question

(1) $\dfrac{2}{4} = \dfrac{\boxed{}}{2}$

(2) $\dfrac{4}{6} = \dfrac{\boxed{}}{3}$

(3) $\dfrac{6}{8} = \dfrac{\boxed{}}{4}$

(4) $\dfrac{3}{6} = \dfrac{\boxed{}}{2}$

(5) $\dfrac{6}{9} = \dfrac{\boxed{}}{3}$

(6) $\dfrac{9}{12} = \dfrac{\boxed{}}{4}$

(7) $\dfrac{4}{8} = \dfrac{\boxed{}}{2}$

(8) $\dfrac{8}{12} = \dfrac{\boxed{}}{3}$

(9) $\dfrac{12}{16} = \dfrac{\boxed{}}{4}$

(10) $\dfrac{5}{10} = \dfrac{\boxed{}}{2}$

(11) $\dfrac{10}{15} = \dfrac{\boxed{}}{3}$

(12) $\dfrac{15}{20} = \dfrac{\boxed{}}{4}$

(13) $\dfrac{6}{12} = \dfrac{\boxed{}}{2}$

(14) $\dfrac{12}{18} = \dfrac{\boxed{}}{3}$

(15) $\dfrac{24}{32} = \dfrac{\boxed{}}{4}$

(16) $\dfrac{18}{27} = \dfrac{\boxed{}}{3}$

(17) $\dfrac{33}{44} = \dfrac{\boxed{}}{4}$

(18) $\dfrac{26}{39} = \dfrac{\boxed{}}{3}$

(19) $\dfrac{12}{8} = \dfrac{\boxed{}}{2}$

(20) $\dfrac{15}{12} = \dfrac{\boxed{}}{4}$

You got it! Good job.

17

1 Reduce the following fractions to simplest form. (An easy way to do this is to find the greatest common factor of the numerator and the denominator and divide each by the GCF.) 2 points per question

(1) $\frac{2}{6} = \frac{}{3}$

(2) $\frac{2}{8} =$

(3) $\frac{2}{12} =$

(4) $\frac{3}{9} = \frac{}{3}$

(5) $\frac{3}{12} =$

(6) $\frac{9}{12} =$

(7) $\frac{4}{8} =$

(8) $\frac{4}{16} =$

(9) $\frac{12}{20} =$

(10) $\frac{5}{10} =$

(11) $\frac{5}{15} =$

(12) $\frac{20}{25} =$

(13) $\frac{6}{12} =$

(14) $\frac{6}{18} =$

(15) $\frac{24}{30} =$

(16) $\frac{7}{14} =$

(17) $\frac{14}{21} =$

(18) $\frac{28}{35} =$

(19) $\frac{8}{16} =$

(20) $\frac{8}{24} =$

(21) $\frac{32}{40} =$

(22) $\frac{9}{18} =$

(23) $\frac{27}{36} =$

(24) $\frac{18}{45} =$

(25) $\frac{10}{8} =$

(26) $\frac{12}{9} =$

(27) $\frac{20}{12} =$

(28) $\frac{18}{12} =$

(29) $\frac{15}{60} =$

(30) $\frac{13}{39} =$

Don't forget!

If the numerator and the denominator of a fraction are both multiplied by the same number, the resulting fraction is equal to the original fraction.

⟨Example⟩ $\frac{2}{3} = \frac{4}{6} = \frac{6}{9}$, $\frac{2}{5} = \frac{4}{10} = \frac{6}{15}$, $\frac{7}{6} = \frac{14}{12} = \frac{21}{18}$

2   Write the appropriate number in each box below.

2 points per question

(1) $\frac{1}{2} = \frac{\boxed{}}{4}$

(2) $\frac{1}{2} = \frac{\boxed{}}{6}$

(3) $\frac{1}{2} = \frac{\boxed{}}{8}$

(4) $\frac{1}{3} = \frac{\boxed{}}{6}$

(5) $\frac{1}{3} = \frac{\boxed{}}{12}$

(6) $\frac{2}{3} = \frac{\boxed{}}{9}$

(7) $\frac{1}{4} = \frac{\boxed{}}{8}$

(8) $\frac{3}{4} = \frac{\boxed{}}{8}$

(9) $\frac{3}{4} = \frac{\boxed{}}{12}$

(10) $\frac{1}{5} = \frac{\boxed{}}{10}$

(11) $\frac{2}{5} = \frac{\boxed{}}{15}$

(12) $\frac{4}{5} = \frac{\boxed{}}{20}$

(13) $\frac{1}{6} = \frac{\boxed{}}{12}$

(14) $\frac{5}{6} = \frac{\boxed{}}{18}$

(15) $\frac{1}{7} = \frac{\boxed{}}{14}$

(16) $\frac{2}{7} = \frac{\boxed{}}{21}$

(17) $\frac{8}{7} = \frac{\boxed{}}{28}$

(18) $\frac{5}{8} = \frac{\boxed{}}{16}$

(19) $\frac{4}{9} = \frac{\boxed{}}{27}$

(20) $\frac{11}{9} = \frac{\boxed{}}{36}$

No worries, right? Excellent.

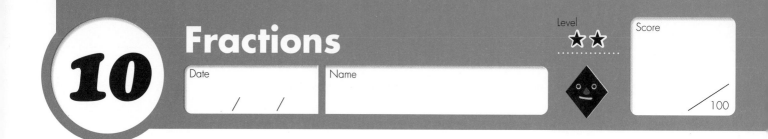

Don't forget!

In order to add two fractions with different denominators, you need to change the fractions so that they both have their least common multiple in the denominator.

〈Example〉 In order to add $\frac{1}{2}$ and $\frac{2}{3}$, you must find their least common multiple in the denominator.

1 First, find the least common denominator for each pair of fractions below. Then change the numerators.

3 points per question

(1) $\left(\frac{1}{4}, \frac{2}{3}\right)$ $\left(\dfrac{\square}{12}, \dfrac{\square}{12}\right)$

(2) $\left(\frac{1}{3}, \frac{3}{4}\right)$ $(\quad , \quad)$

(3) $\left(\frac{1}{4}, \frac{2}{5}\right)$ $(\quad , \quad)$

(4) $\left(\frac{2}{5}, \frac{1}{6}\right)$ $(\quad , \quad)$

(5) $\left(\frac{3}{4}, \frac{5}{6}\right)$ $\left(\dfrac{\square}{12}, \dfrac{\square}{12}\right)$

(6) $\left(\frac{1}{6}, \frac{2}{9}\right)$ $(\quad , \quad)$

(7) $\left(\frac{3}{8}, \frac{5}{12}\right)$ $(\quad , \quad)$

(8) $\left(\frac{3}{5}, \frac{7}{15}\right)$ $(\quad , \quad)$

(9) $\left(\frac{3}{2}, \frac{4}{3}\right)$ $(\quad , \quad)$

(10) $\left(\frac{7}{5}, \frac{5}{4}\right)$ $(\quad , \quad)$

(11) $\left(\frac{5}{3}, \frac{19}{12}\right)$ $(\quad , \quad)$

(12) $\left(\frac{7}{6}, \frac{9}{8}\right)$ $(\quad , \quad)$

2 Change each pair of fractions below so that they have the same denominator. Then circle the larger original number.

4 points per question

(1) $\left(\frac{1}{2}, \frac{2}{3}\right)$ $\left(\frac{}{6}, \frac{}{6}\right)$

(2) $\left(\frac{2}{3}, \frac{3}{4}\right)$ $\left(\quad, \quad\right)$

(3) $\left(\frac{3}{4}, \frac{4}{5}\right)$ $\left(\quad, \quad\right)$

(4) $\left(\frac{6}{7}, \frac{7}{8}\right)$ $\left(\quad, \quad\right)$

(5) $\left(\frac{3}{5}, \frac{1}{2}\right)$ $\left(\quad, \quad\right)$

(6) $\left(\frac{1}{3}, \frac{4}{9}\right)$ $\left(\quad, \quad\right)$

(7) $\left(\frac{7}{12}, \frac{3}{4}\right)$ $\left(\quad, \quad\right)$

(8) $\left(\frac{3}{5}, \frac{11}{15}\right)$ $\left(\quad, \quad\right)$

(9) $\left(\frac{5}{8}, \frac{7}{12}\right)$ $\left(\quad, \quad\right)$

(10) $\left(\frac{8}{9}, \frac{5}{6}\right)$ $\left(\quad, \quad\right)$

(11) $\left(\frac{3}{4}, \frac{4}{3}\right)$ $\left(\quad, \quad\right)$

(12) $\left(\frac{6}{5}, \frac{5}{4}\right)$ $\left(\quad, \quad\right)$

(13) $\left(\frac{3}{2}, \frac{5}{3}\right)$ $\left(\quad, \quad\right)$

(14) $\left(\frac{11}{6}, \frac{9}{5}\right)$ $\left(\quad, \quad\right)$

(15) $\left(\frac{9}{8}, \frac{17}{16}\right)$ $\left(\quad, \quad\right)$

(16) $\left(\frac{5}{4}, \frac{7}{6}\right)$ $\left(\quad, \quad\right)$

A little more work with fractions, okay?

1 Rewrite the measurements below in fractions of hours.

2 points per question

(1) 60 minutes = ☐ hour

(2) 30 minutes = $\frac{30}{60}$ hour = ☐ hour

(3) 15 minutes = $\frac{15}{60}$ hour = ☐ hour

(4) 20 minutes = $\frac{20}{60}$ hour = ☐ hour

(5) 10 minutes = ☐ hour

(6) 5 minutes = ☐ hour

(7) 1 minute = ☐ hour

(8) 45 minutes = ☐ hour

(9) 1 hour 50 minutes = $1\frac{50}{60}$ hours = ☐ hours

(10) 3 hours 25 minutes = ☐ hours

2 Rewrite the measurements below in minutes.

2 points per question

(1) 1 hour = ☐ minutes

(2) $\frac{1}{60}$ hour = $\left(60 \times \frac{1}{60}\right)$ minute = ☐ minute

(3) $\frac{17}{60}$ hour = $\left(60 \times \frac{17}{60}\right)$ minutes = ☐ minutes

(4) $\frac{1}{30}$ hour = $\left(60 \times \frac{1}{30}\right)$ minutes = ☐ minutes

(5) $\frac{1}{20}$ hour = $\left(60 \times \frac{1}{20}\right)$ minutes = ☐ minutes

(6) $\frac{1}{15}$ hour = $\left(60 \times \frac{1}{15}\right)$ minutes = ☐ minutes

(7) $\frac{1}{10}$ hour = ☐ minutes

(8) $\frac{1}{6}$ hour = ☐ minutes

(9) $\frac{5}{6}$ hour = ☐ minutes

(10) $\frac{1}{5}$ hour = ☐ minutes

(11) $\frac{4}{5}$ hour = ☐ minutes

(12) $\frac{1}{4}$ hour = ☐ minutes

(13) $\frac{3}{4}$ hour = ☐ minutes

(14) $\frac{1}{3}$ hour = ☐ minutes

(15) $8\frac{2}{3}$ hours = ☐ hours ☐ minutes

3 Write the appropriate number in each box below.

2 points per question

(1) 60 seconds = ☐ minute

(2) 30 seconds = $\frac{30}{60}$ minute = ☐ minute

(3) 15 seconds = $\frac{15}{60}$ minute = ☐ minute

(4) 20 seconds = $\frac{20}{60}$ minute = ☐ minute

(5) 10 seconds = ☐ minute

(6) 5 seconds = ☐ minute

(7) 1 second = ☐ minute

(8) 45 seconds = ☐ minute

(9) 1 minute 12 seconds = $1\frac{12}{60}$ minutes = ☐ minutes

(10) 5 minutes 8 seconds = ☐ minutes

4 Write the appropriate number in each box below.

2 points per question

(1) 1 minute = ☐ seconds

(2) $\frac{1}{60}$ minute = $\left(60 \times \frac{1}{60}\right)$ second = ☐ second

(3) $\frac{59}{60}$ minute = $\left(60 \times \frac{59}{60}\right)$ seconds = ☐ seconds

(4) $\frac{1}{30}$ minute = $\left(60 \times \frac{1}{30}\right)$ seconds = ☐ seconds

(5) $\frac{1}{20}$ minute = $\left(60 \times \frac{1}{20}\right)$ seconds = ☐ seconds

(6) $\frac{1}{15}$ minute = $\left(60 \times \frac{1}{15}\right)$ seconds = ☐ seconds

(7) $\frac{7}{15}$ minute = ☐ seconds

(8) $\frac{1}{10}$ minute = ☐ seconds

(9) $\frac{1}{6}$ minute = ☐ seconds

(10) $\frac{1}{5}$ minute = ☐ seconds

(11) $\frac{4}{5}$ minute = ☐ seconds

(12) $\frac{1}{4}$ minute = ☐ seconds

(13) $\frac{3}{4}$ minute = ☐ seconds

(14) $\frac{1}{3}$ minute = ☐ seconds

(15) $4\frac{2}{3}$ minutes = ☐ minutes ☐ seconds

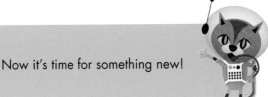

Now it's time for something new!

Circles

Don't forget!

The perimeter of a circle is called the **circumference**.

The length of the circumference is equal to the diameter times pi, which is written π, and has an estimated value of 3.14. π is the ratio of a circle's circumference to its diameter.

circumference
(circumference = diameter $\times \pi$)

diameter
(diameter = radius $\times$ 2)

---

1 What is the circumference of each circle below? Remember to use $\pi = 3.14$.

6 points per question

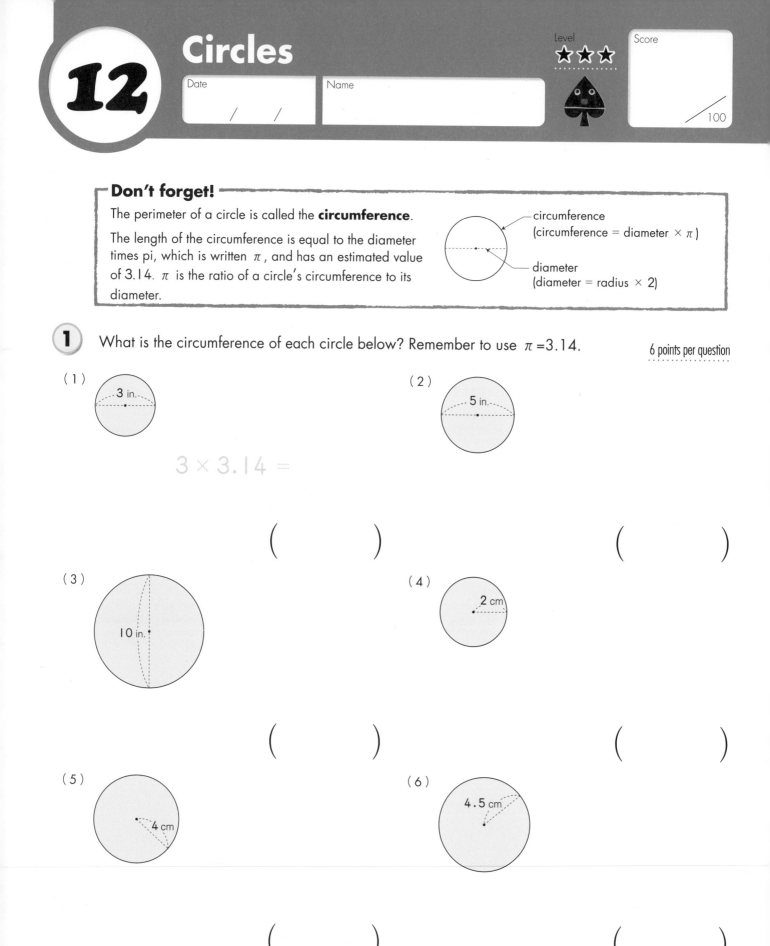

(1)

3 in.

$3 \times 3.14 =$

( )

(2)

5 in.

( )

(3)

10 in.

( )

(4)

2 cm

( )

(5)

4 cm

( )

(6)

4.5 cm

( )

2 What is the perimeter of each shape below?

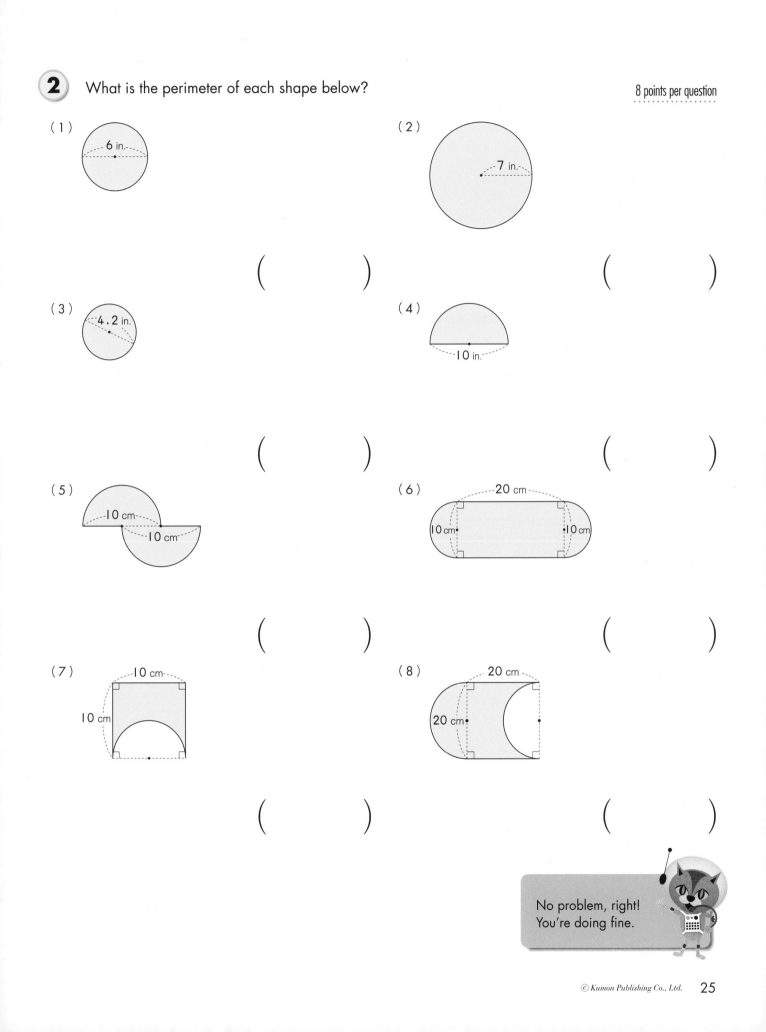

(1)

6 in.

(   )

(2)

7 in.

(   )

(3)

4.2 in.

(   )

(4)

10 in.

(   )

(5)

10 cm
10 cm

(   )

(6)

20 cm

10 cm   10 cm

(   )

(7)

10 cm
10 cm

(   )

(8)

20 cm
20 cm

(   )

No problem, right!
You're doing fine.

Circles

Don't forget!

The area of a circle is equal to the radius squared times π (3.14).

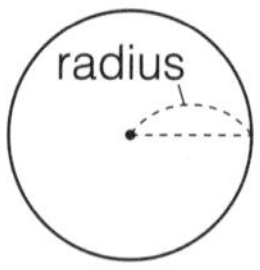

radius

1 What is the area of each circle below? Remember to use π =3.14.

6 points per question

(1)

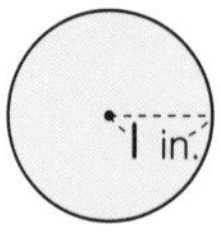

1 in.

$1 \times 1 \times 3.14 =$

(   )

(2)

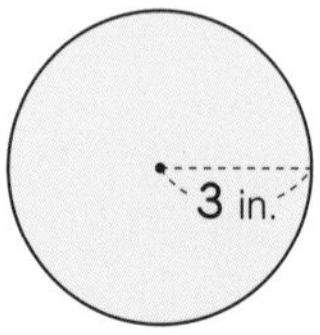

3 in.

(   )

(3)

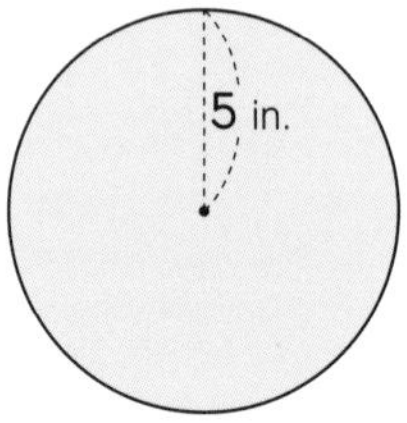

5 in.

(   )

(4)

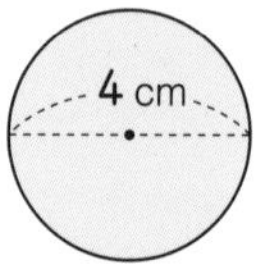

4 cm

(   )

(5)

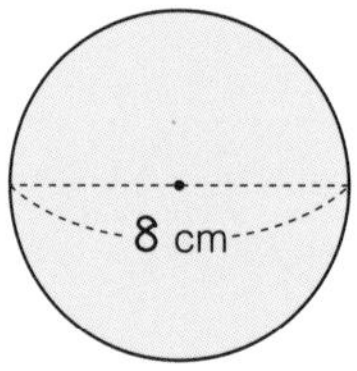

8 cm

(   )

(6)

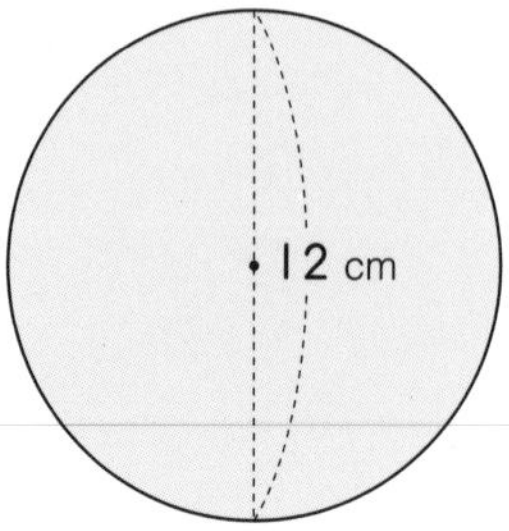

12 cm

(   )

2   What is the area of each shape below?

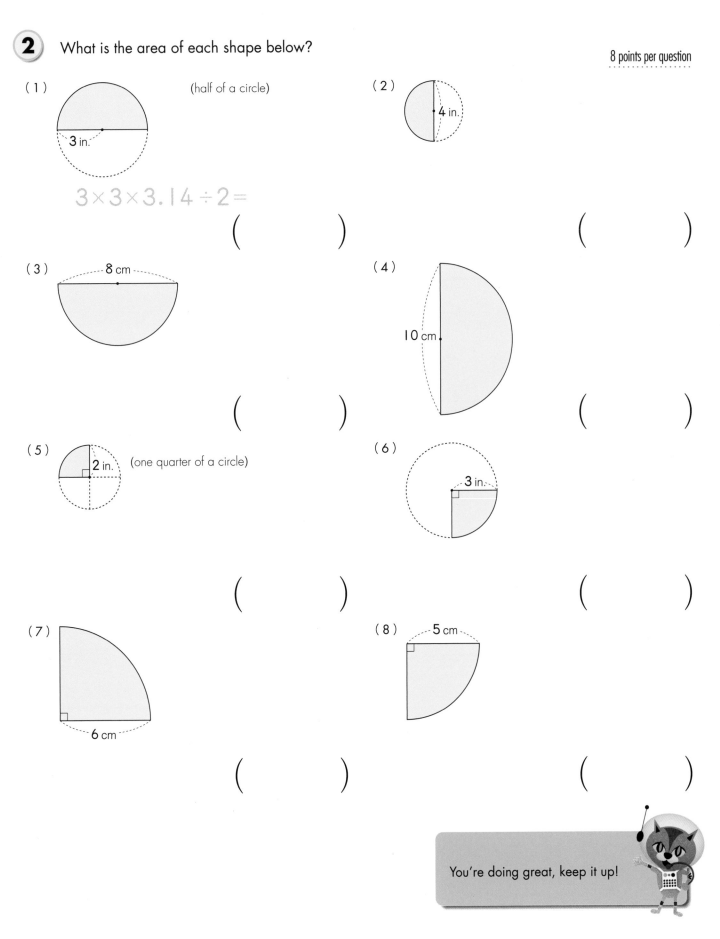

(1)   (half of a circle)

3 in.

$3 \times 3 \times 3.14 \div 2 =$

(   )

(2)   4 in.

(   )

(3)   8 cm

(   )

(4)   10 cm

(   )

(5)   2 in.   (one quarter of a circle)

(   )

(6)   3 in.

(   )

(7)   6 cm

(   )

(8)   5 cm

(   )

You're doing great, keep it up!

14

Circles

Date   /   /

Name

Level
★★★

Score
/100

1 What is the area of the shaded region in each shape below? Remember to use $\pi = 3.14$.

8 points per question

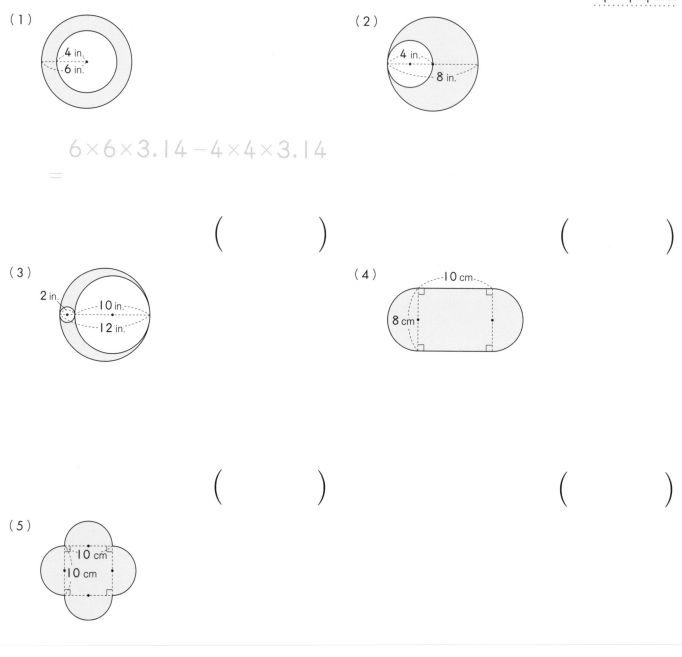

(1)

4 in.
6 in.

$6 \times 6 \times 3.14 - 4 \times 4 \times 3.14$
$=$

(   )

(2)

4 in.
8 in.

(   )

(3)

2 in.
10 in.
12 in.

(   )

(4)

10 cm
8 cm

(   )

(5)

10 cm
10 cm

(   )

2   What is the area of the shaded region in each shape below?

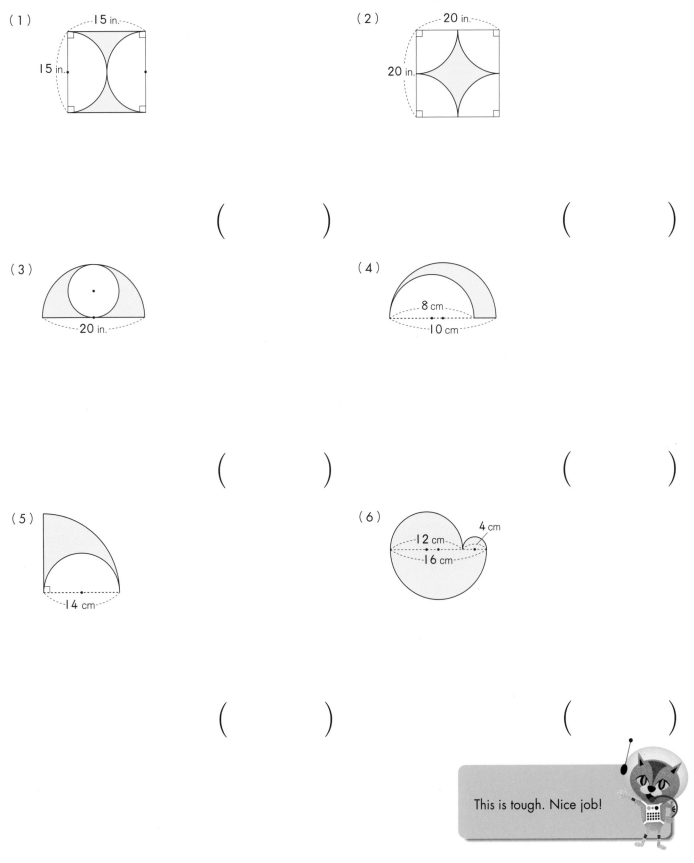

(1)

15 in.

15 in.

(   )

(2)

20 in.

20 in.

(   )

(3)

20 in.

(   )

(4)

8 cm

10 cm

(   )

(5)

14 cm

(   )

(6)

4 cm

12 cm

16 cm

(   )

This is tough. Nice job!

   29

Level ★★

Score
/100

Date / /

Name

Don't forget!

A **sector** is a region bound by an arc and two radii of a circle.

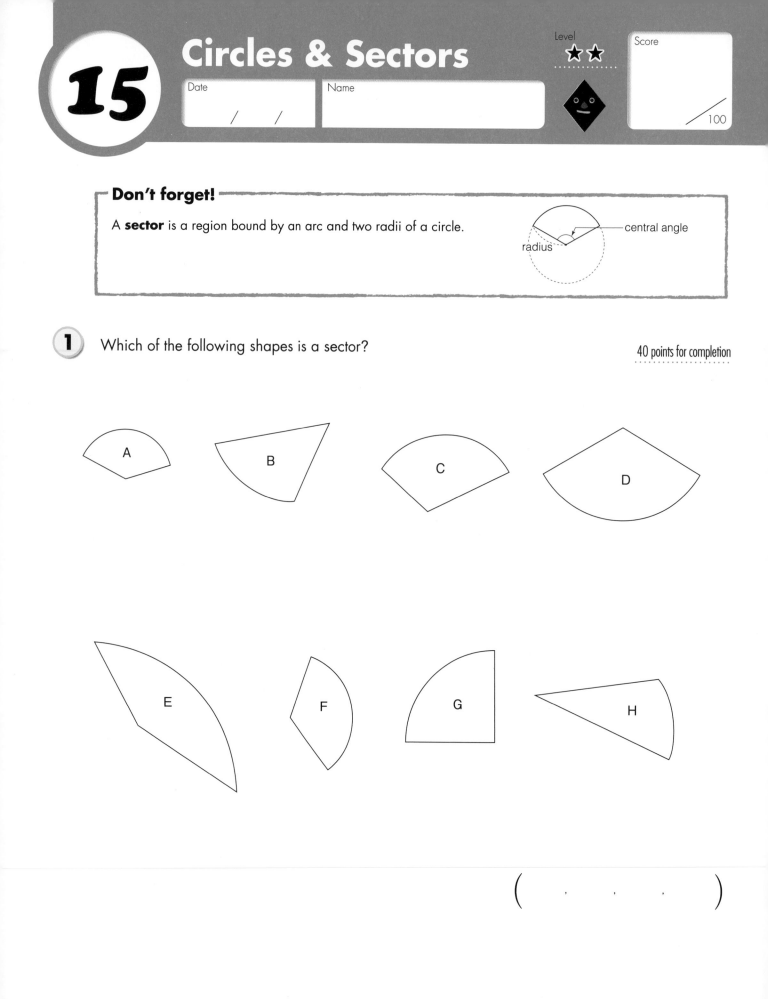

central angle

radius

---

1 Which of the following shapes is a sector?

40 points for completion

A

B

C

D

E

F

G

H

(, , ,)

2   Draw each sector below.

15 points per question

（1）The radius is **2** centimeters long and the central angle is 40°.

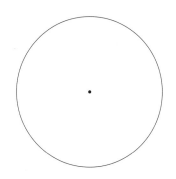

（2）The radius is **3** centimeters long and the central angle is 80°.

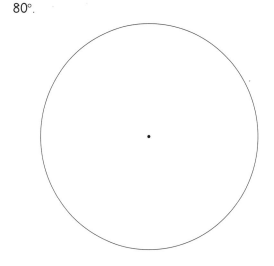

（3）The radius is **2** centimeters long and the central angle is 120°.

（4）The radius is **3** centimeters long and the central angle is 200°.

This is fun, right?

16 **Circles & Sectors**

Date   /   /

Name

Level ★★

Score   /100

---

1 The circle pictured on the right has a radius of 8 centimeters. Answer the questions below about this circle.

8 points per question

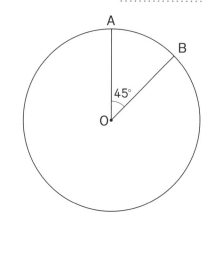

(1) What is the area of the circle?

(   )

(2) What fraction of the area of the circle is the area of the sector AOB?

$\dfrac{45}{360} =$ (   )

(3) What is the area of the sector AOB?

(   )

Don't forget!

The area of a sector $=$ radius$^2 \times \pi \times \dfrac{a}{360}$ (*a* is the central angle of the sector)

---

2 What is the area of each sector below?

8 points per question

(1) Radius = **3** inches   Central angle = **60°**

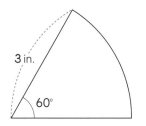

(2) Radius = **9** inches   Central angle = **150°**

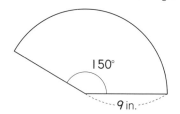

(   )   (   )

32   © Kumon Publishing Co., Ltd.

3 The circle pictured on the right has a radius of 3 inches. Answer the questions below about this circle.

10 points per question

(1) What fraction of the circumference is the length of curve **AB**?

(   )

(2) Find the length of curve **AB**.

(   )

(3) What fraction of the circumference is the length of curve **BC**?

(   )

(4) What fraction of the circumference is the length of curve **CD**?

(   )

(5) Find the perimeter of sector **AOB**.

(   )

(6) Find the perimeter of sector **BOC**.

(   )

┌ Don't forget! ─────────────────────────────

The length of the curve in a sector

$$= 2 \times \pi \times radius \times \frac{a}{360} \text{ (}a \text{ is central angle)}$$

The perimeter of a sector

$$= 2 \times \pi \times radius \times \frac{a}{360} + 2 \times radius$$

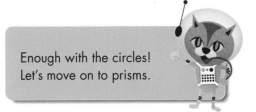

Enough with the circles! Let's move on to prisms.

33

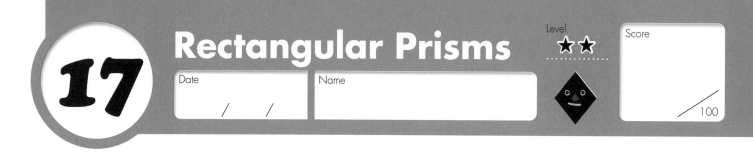

Rectangular Prisms

Date   /   /

Name

Don't forget!

A figure made up of only rectangles or squares is called a **rectangular prism**.

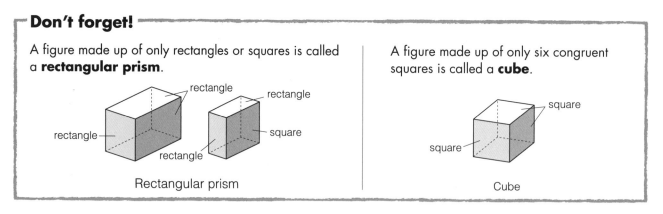

rectangle   rectangle

rectangle   square

rectangle

rectangle

Rectangular prism

A figure made up of only six congruent squares is called a **cube**.

square

square

Cube

1   Flat faces are called plane surfaces. The figures below are bound by different plane surfaces. Judging by their plane surfaces, which of the shapes are rectangular prisms? Which ones are cubes?

8 points for completion

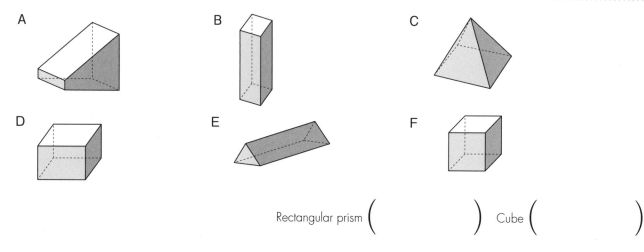

A   B   C

D   E   F

Rectangular prism (   )   Cube (   )

2   Fill the table on the right with the number of faces, edges and vertices of the rectangular prism and cube pictured below.

6 points per box

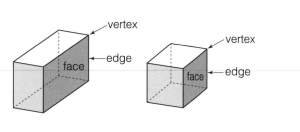

vertex

edge

face

vertex

face   edge

	Rectangular prism	Cube
Number of faces		
Number of edges		
Number of vertices		

34   © Kumon Publishing Co., Ltd.

3 The rectangular prism on the right has edges of 6 centimeters, 8 centimeters, and 3 centimeters.

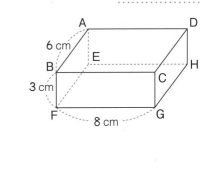

(1) Which edges are equal to edge **AB**?

$\Big($edge $\Big)$, $\Big($edge $\Big)$, $\Big($edge $\Big)$

(2) Which edges are equal to edge **AD**?

$\Big($edge $\Big)$, $\Big($edge $\Big)$, $\Big($edge $\Big)$

(3) Which edges are equal to edge **AE**?

$\Big($edge $\Big)$, $\Big($edge $\Big)$, $\Big($edge $\Big)$

(4) How many faces have a length of **3** centimeters and a width of **6** centimeters?

$\Big($ $\Big)$

(5) How many faces have a length of **3** centimeters and a width of **8** centimeters?

$\Big($ $\Big)$

(6) How many faces have a length of **6** centimeters and a width of **8** centimeters?

$\Big($ $\Big)$

4 The cube on the right has edges that are 4 centimeters long.

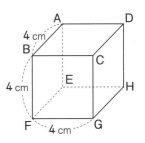

(1) Which edges are equal to edge **AB**?

(2) How many faces have a length and a width of **4** centimeters?

$\Big($ $\Big)$

See? Rectangular prisms and cubes are cool.

18 Rectangular Prisms

Level ★★

Date / /

Name

Score /100

1 Answer the questions below about the rectangular prism on the right.

8 points per question

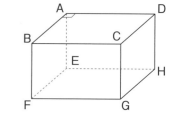

（1） Edge **AB** and edge **AD** are perpendicular lines. Name the edges that are perpendicular to edge **AB** through vertex **A**.

$$\left(\text{edge} \quad AD \quad \right), \left(\text{edge} \qquad \right)$$

（2） Which edges are perpendicular to edge **BF**?

$$\left(\text{edge} \qquad \right), \left(\text{edge} \qquad \right), \left(\text{edge} \qquad \right), \left(\text{edge} \qquad \right)$$

（3） Edges **AB** and **DC** are parallel. Which edges are parallel to edge **AB**?

$$\left(\text{edge} \quad DC \quad \right), \left(\text{edge} \qquad \right), \left(\text{edge} \qquad \right)$$

（4） Which edges are parallel to edge **BC**?

$$\left(\text{edge} \qquad \right), \left(\text{edge} \qquad \right), \left(\text{edge} \qquad \right)$$

2 Answer the questions below about the rectangular prism on the right.

8 points per question

（1） Edge **AE** intersects face *a* perpendicularly. Which edges intersect face *a* perpendicularly?

$$\left(\text{edge} \quad AE \quad \right), \left(\text{edge} \qquad \right), \left(\text{edge} \qquad \right), \left(\text{edge} \qquad \right)$$

（2） Which edges intersect face **BCGF** perpendicularly?

$$\left(\text{edge} \qquad \right), \left(\text{edge} \qquad \right), \left(\text{edge} \qquad \right), \left(\text{edge} \qquad \right)$$

36   © Kumon Publishing Co., Ltd.

3 Answer the questions below about the rectangular prism on the right.

8 points per question

(1) Face **AEFB** intersects face *a* perpendicularly. Which other faces intersect face *a* perpendicularly?

$$\left(\text{face} \qquad\right), \left(\text{face} \qquad\right),$$

$$\left(\text{face} \qquad\right)$$

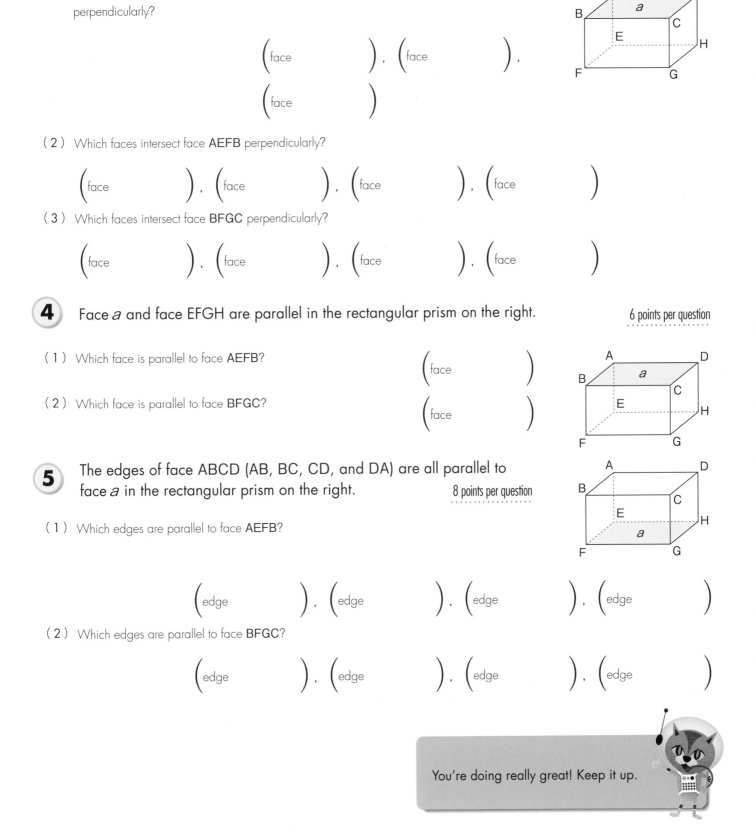

(2) Which faces intersect face **AEFB** perpendicularly?

$$\left(\text{face} \qquad\right), \left(\text{face} \qquad\right), \left(\text{face} \qquad\right), \left(\text{face} \qquad\right)$$

(3) Which faces intersect face **BFGC** perpendicularly?

$$\left(\text{face} \qquad\right), \left(\text{face} \qquad\right), \left(\text{face} \qquad\right), \left(\text{face} \qquad\right)$$

4 Face *a* and face **EFGH** are parallel in the rectangular prism on the right.

6 points per question

(1) Which face is parallel to face **AEFB**?

$$\left(\text{face} \qquad\right)$$

(2) Which face is parallel to face **BFGC**?

$$\left(\text{face} \qquad\right)$$

5 The edges of face **ABCD** (AB, BC, CD, and DA) are all parallel to face *a* in the rectangular prism on the right.

8 points per question

(1) Which edges are parallel to face **AEFB**?

$$\left(\text{edge} \qquad\right), \left(\text{edge} \qquad\right), \left(\text{edge} \qquad\right), \left(\text{edge} \qquad\right)$$

(2) Which edges are parallel to face **BFGC**?

$$\left(\text{edge} \qquad\right), \left(\text{edge} \qquad\right), \left(\text{edge} \qquad\right), \left(\text{edge} \qquad\right)$$

You're doing really great! Keep it up.

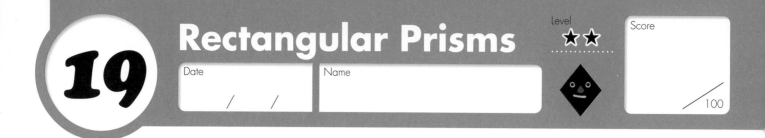

19 Rectangular Prisms

Level ★★

Date   /   /

Name

Score

/100

1 The shapes below have been built by cutting paper and folding it together. Which plans would make the shapes on the left? Write a ✓ under the correct plan.

8 points per question

(1) (Built shape)

(Plan)

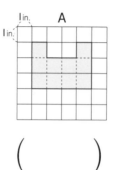

(   )   (   )

(2) (Built shape)

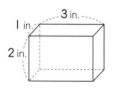

(Plan)

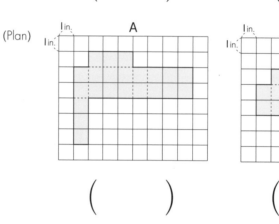

(   )   (   )

2 Answer the questions about the plans below.

9 points per question

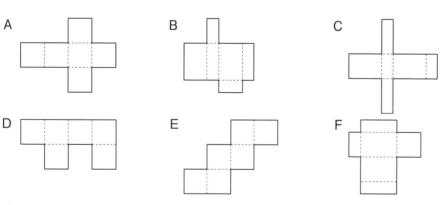

(1) Which plans would build rectangular prisms?   (   )

(2) Which plans would build cubes?   (   )

3 The figures on the right show the plans and the built shape for a rectangle with edges that are 12 centimeters, 8 centimeters and 5 centimeters long.

6 points per question

(Plan)

(Built shape)

8 cm
12 cm
5 cm

A N M ... L K J I
B C D G H
E F

(1) How long is edge **AB**?

(   )

(2) How long is edge **FG**?

(   )

(3) How long is edge **DG**?

(   )

(4) When you fold up the plans, which edge touches edge **AB**?

(edge   )

(5) Which edge touches edge **DE**?

(edge   )

(6) Which edge touches edge **GH**?

(edge   )

(7) When you fold up the plans, which vertex touches vertex **C**?

(vertex   )

(8) Which vertex touches vertex **L**?

(vertex   )

(9) Which vertex touches vertex **A**?

(vertex   ) , (vertex   )

(10) When you fold up the plans, which face is parallel to face **KJML**?

(face   )

(11) When you fold up the plans, which faces intersect face **NMDC** perpendicularly?

(   )

You may want to get out some graph paper and try to make your own plans for a shape.

Rectangular Prisms

Date   /   /

Name

Score

/100

1 Draw a plan that would make each shape on the left.

14 points per question

(1)

(2)

(3)

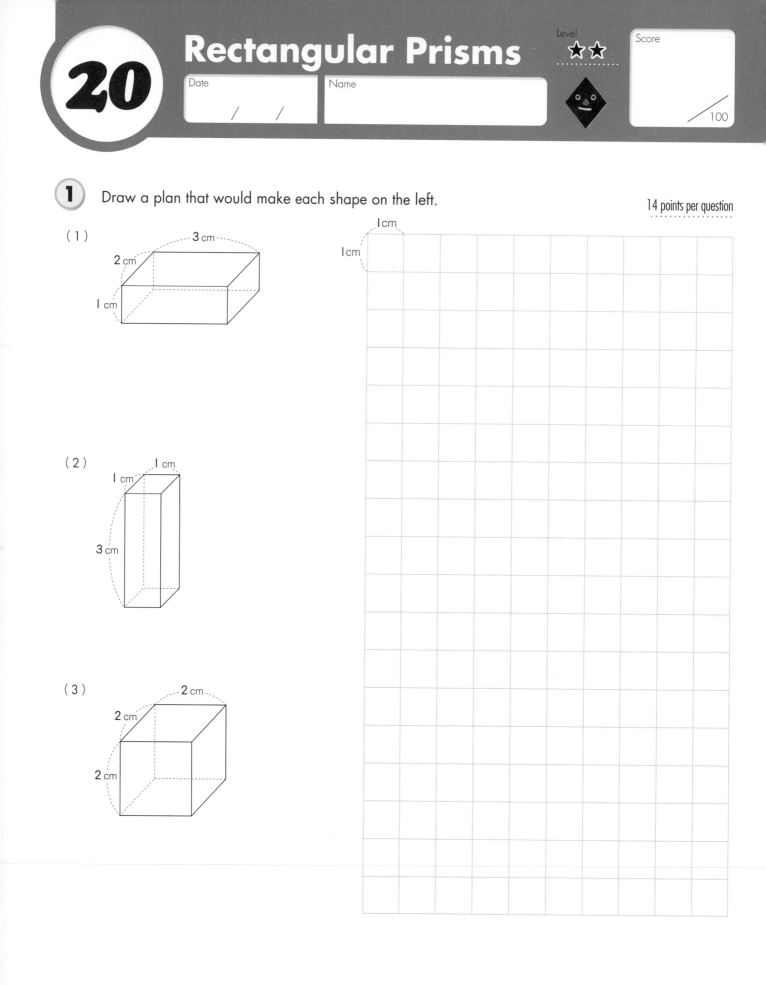

2   Draw the shape that would be made by each plan on the left.

14 points per question

（1）rectangular prism

（2）cube

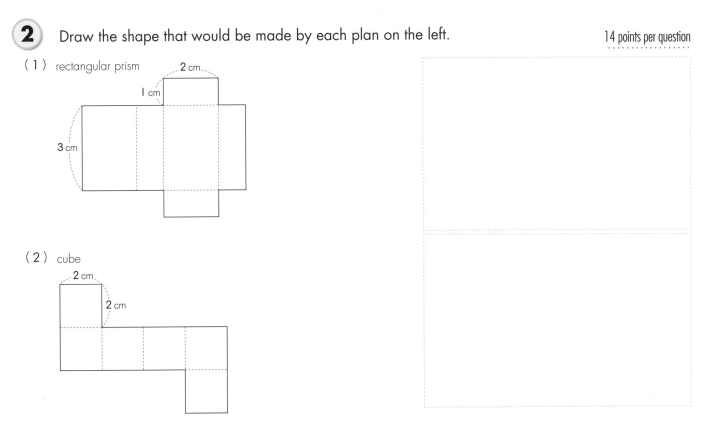

3   Draw the shape and the plan that would make a rectangular prism with edges that are 2 centimeters, 3 centimeters and 2 centimeters long.

30 points for completion

(Built shape)

(Plan)

Try to visualize how your plan would fold.
Well done!

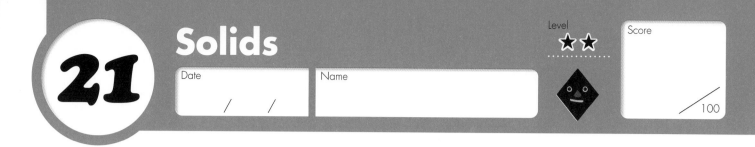

21 Solids

Date   /   /

Name

Level ★★

Score   /100

Don't forget!

The solids below are called **prisms**.

Triangular prism   Quadrilateral prism   Pentagonal prism

triangle   quadrilateral   pentagon

The solids below are called **cylinders**.

Cylinder   Cylinder

circle

1 What is the name of each solid below?

4 points per question

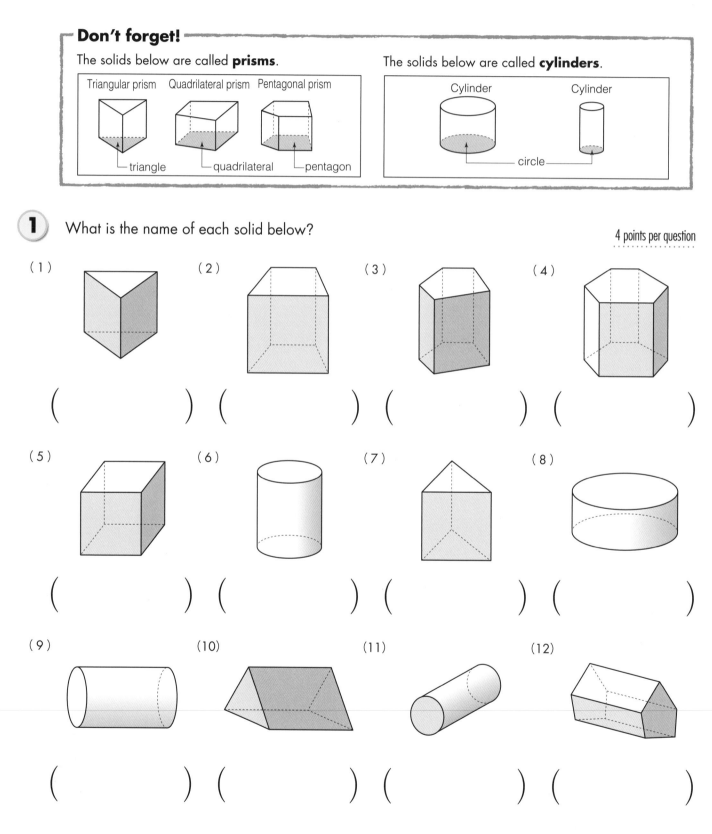

(1)   (2)   (3)   (4)

(   ) (   ) (   ) (   )

(5)   (6)   (7)   (8)

(   ) (   ) (   ) (   )

(9)   (10)   (11)   (12)

(   ) (   ) (   ) (   )

Don't forget!

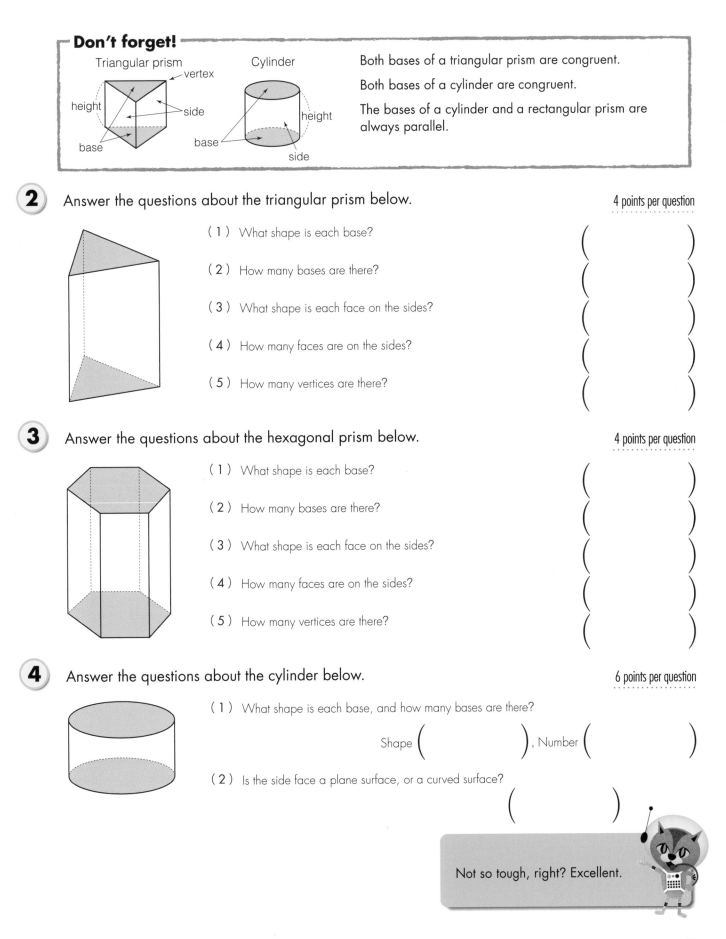

Triangular prism

vertex

height

side

base

Cylinder

height

base

side

Both bases of a triangular prism are congruent.

Both bases of a cylinder are congruent.

The bases of a cylinder and a rectangular prism are always parallel.

2 Answer the questions about the triangular prism below.

4 points per question

(1) What shape is each base?   (   )

(2) How many bases are there?   (   )

(3) What shape is each face on the sides?   (   )

(4) How many faces are on the sides?   (   )

(5) How many vertices are there?   (   )

3 Answer the questions about the hexagonal prism below.

4 points per question

(1) What shape is each base?   (   )

(2) How many bases are there?   (   )

(3) What shape is each face on the sides?   (   )

(4) How many faces are on the sides?   (   )

(5) How many vertices are there?   (   )

4 Answer the questions about the cylinder below.

6 points per question

(1) What shape is each base, and how many bases are there?

Shape (   ) , Number (   )

(2) Is the side face a plane surface, or a curved surface?

(   )

Not so tough, right? Excellent.

22 Solids

Date   /   /

Name

Level
★★★

Score
/100

1 Which of the following plans would build a triangular prism?   40 points for completion

A
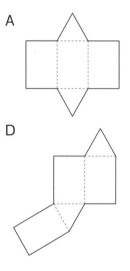

B

C

D

E
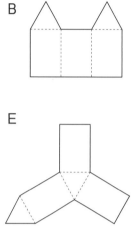

F
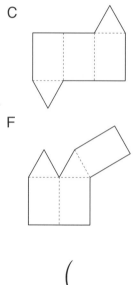

(   )

2 Below are the plans and the built shape of a cylinder.   5 points per question

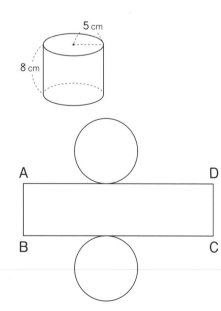

5 cm

8 cm

A   D

B   C

(1) What face in the cylinder does the rectangle **ABCD** represent?

(   )

(2) The length of the side **AD** equal to the circle's _____?

(   )

(3) How long are sides **AB** and **AD**?

AB (   )   AD (   )

(4) How tall is the cylinder?

(   )

3 Draw a plan for making the triangular prism below.

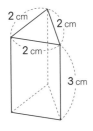

4 Draw a plan for the cylinder below.

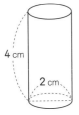

Remember you can always try these for yourself with some graph paper.

23 Solids

Level ★★★

Date / /

Name

Score /100

Don't forget!

The solids below are called **pyramids**.

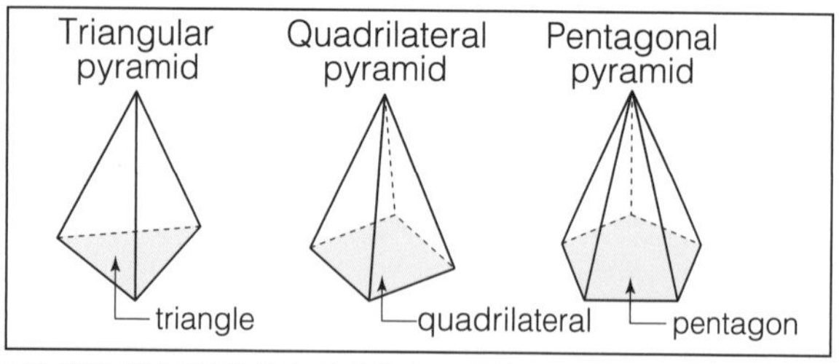

Triangular pyramid
Quadrilateral pyramid
Pentagonal pyramid
triangle
quadrilateral
pentagon

The solids below are called **cones**.

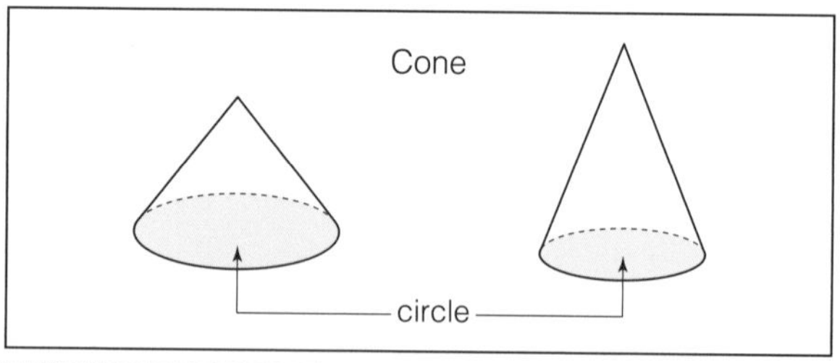

Cone
circle

1 What is the name of each solid below?

8 points per question

(1)

( )

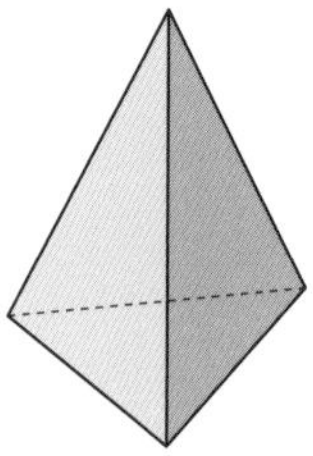

(2)

( )

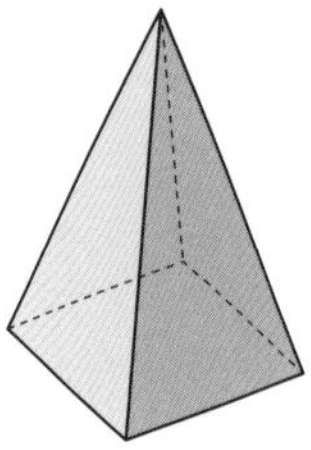

(3)

( )

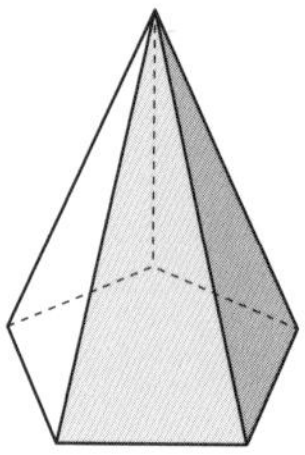

(4)

( )

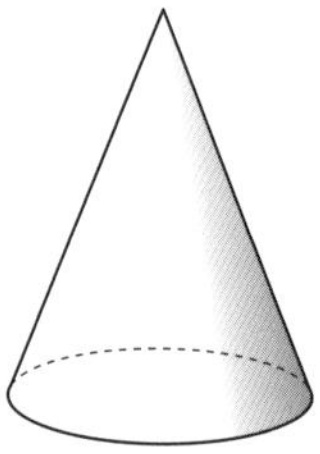

Don't forget!

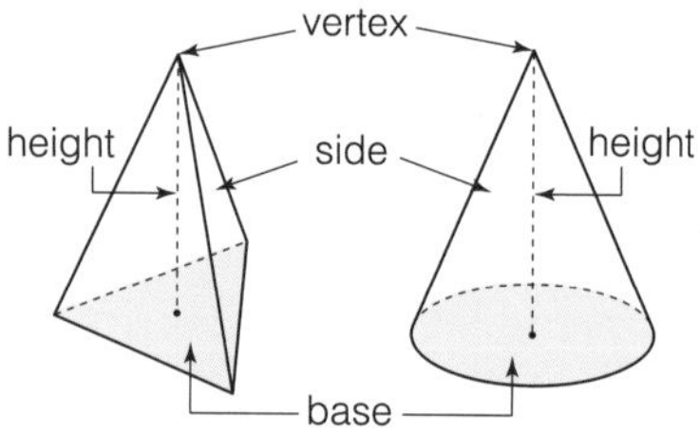

vertex
height
side
height
base

The base of a pyramid is a polygon.
The side faces of a pyramid are isosceles triangles.

The base of a cone is a circle.
The side face of a cone is a curved surface.

2 Answer the questions about the quadrilateral pyramid below.

8 points per question

4 cm
4 cm
4 cm
4 cm

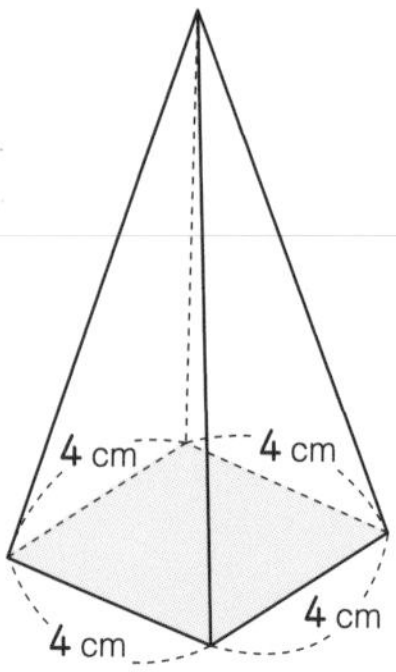

(1) What shape is the base? ( )

(2) How many bases are there? ( )

(3) What shape is each side face? ( )

(4) How many side faces are there? ( )

3 Answer the questions about the cone below.

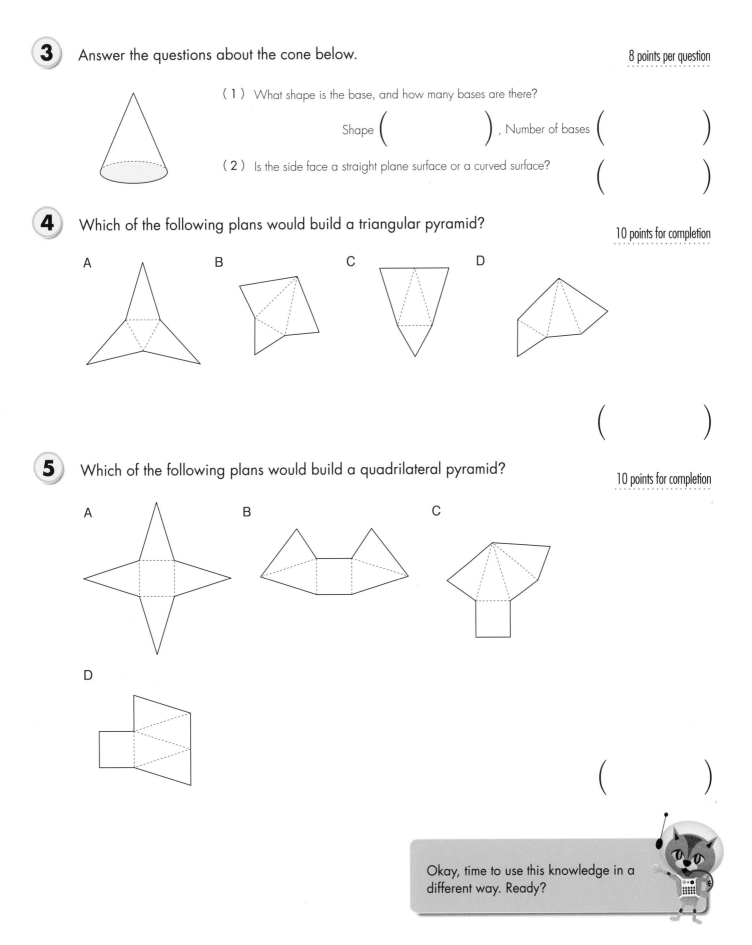

(1) What shape is the base, and how many bases are there?

Shape (   ) , Number of bases (   )

(2) Is the side face a straight plane surface or a curved surface?

(   )

4 Which of the following plans would build a triangular pyramid?

10 points for completion

A   B   C   D

(   )

5 Which of the following plans would build a quadrilateral pyramid?

10 points for completion

A   B   C

D

(   )

Okay, time to use this knowledge in a different way. Ready?

Surface Area & Volume

Date / /

Name

Score /100

Don't forget!

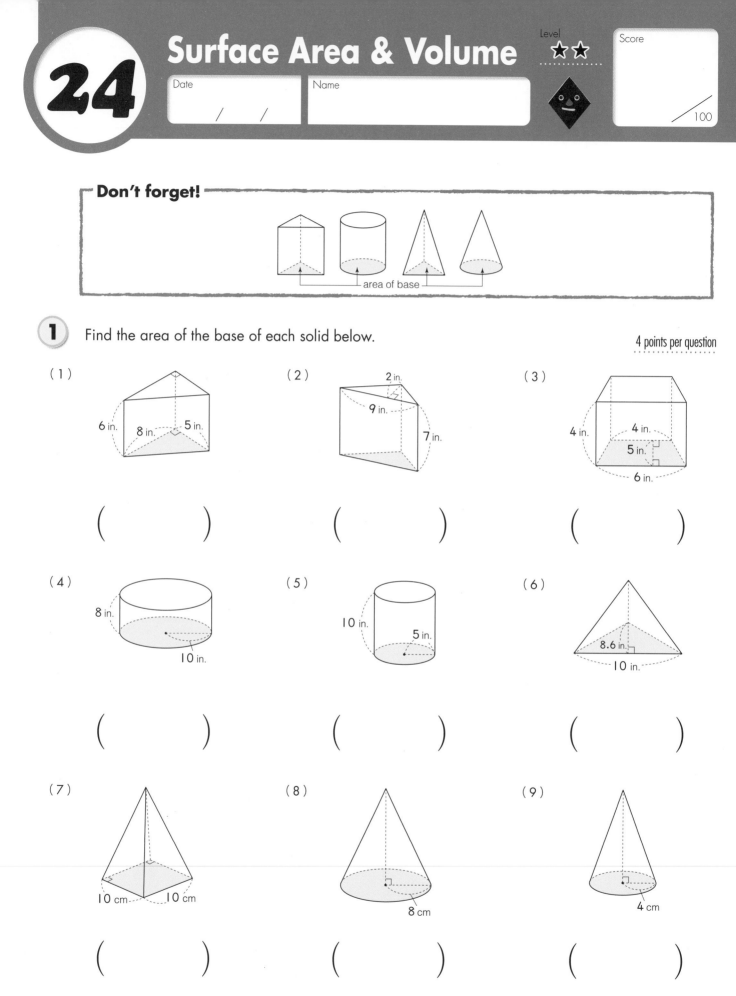

area of base

1 Find the area of the base of each solid below.

4 points per question

(1)

6 in.   8 in.   5 in.

(   )

(2)

2 in.
9 in.
7 in.

(   )

(3)

4 in.   4 in.
5 in.
6 in.

(   )

(4)

8 in.
10 in.

(   )

(5)

10 in.
5 in.

(   )

(6)

8.6 in.
10 in.

(   )

(7)

10 cm   10 cm

(   )

(8)

8 cm

(   )

(9)

4 cm

(   )

Don't forget!

Volume of prism or cylinder = area of base × height

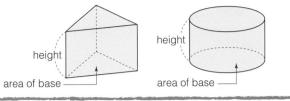

height
area of base
height
area of base

2 Find the volume of each solid below.

8 points per question

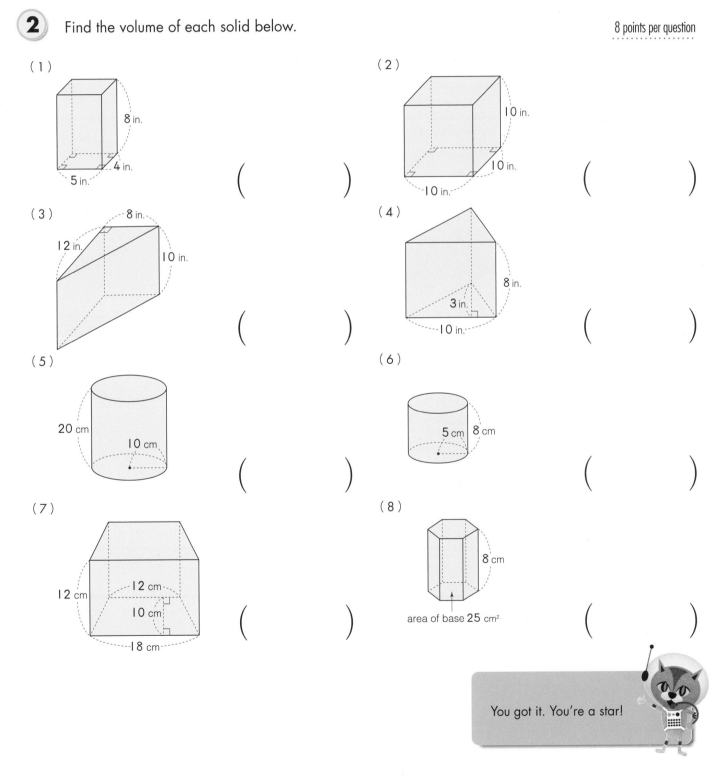

(1)
8 in.
4 in.
5 in.

( )

(2)
10 in.
10 in.
10 in.

( )

(3)
8 in.
12 in.
10 in.

( )

(4)
8 in.
3 in.
10 in.

( )

(5)
20 cm
10 cm

( )

(6)
5 cm 8 cm

( )

(7)
12 cm
12 cm
10 cm
18 cm

( )

(8)
8 cm
area of base 25 cm²

( )

You got it. You're a star!

49

Surface Area & Volume

Date   /   /

Name

Score   /100

Don't forget!

Volume of a pyramid or a cone = area of base × height × $\frac{1}{3}$

height

area of base

---

1 Find the volume of each pyramid below.

6 points per question

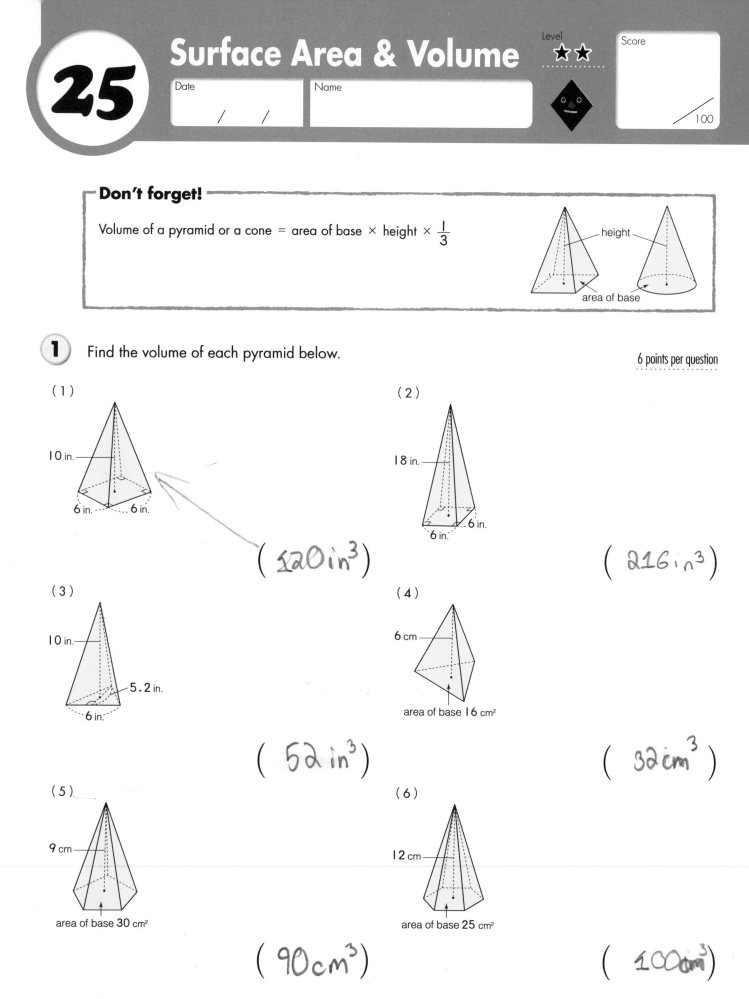

(1)

10 in.

6 in.   6 in.

(120 in³)

(2)

18 in.

6 in.

6 in.

(216 in³)

(3)

10 in.

5.2 in.

6 in.

(52 in³)

(4)

6 cm

area of base 16 cm²

(32 cm³)

(5)

9 cm

area of base 30 cm²

(90 cm³)

(6)

12 cm

area of base 25 cm²

(100 cm³)

2 Find the volume of each cone below.

(1)

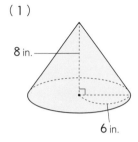

8 in.

6 in.

(301.44 in^3)

(2)

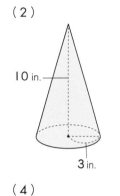

10 in.

3 in.

(94.2 in^3)

(3)

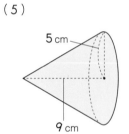

15 in.

10 in.

(1570 in^3)

(4)

6 in.

9 in.

(508.68 in^3)

(5)

5 cm

9 cm

(235.5 cm^3)

(6)

15 cm

4 cm

(251.2 cm^3)

(7)

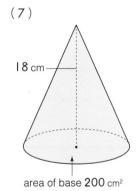

18 cm

area of base 200 cm²

(1200 cm^3)

(8)

11 cm

area of base 18 cm²

(66 cm^3)

I wonder if you could figure out the volume of the pyramids in Egypt!

Don't forget!

The **lateral area** of a solid is the area of all the faces of a solid (prisms, cylinders, pyramids and cones).

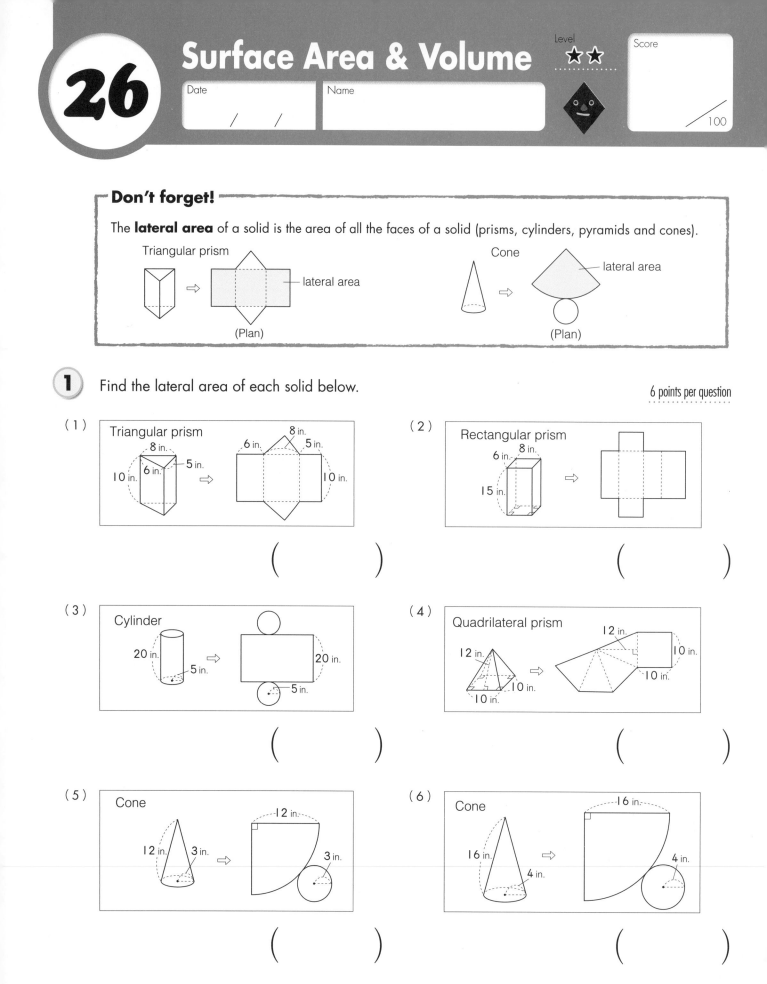

Triangular prism ⇒ lateral area (Plan)

Cone ⇒ lateral area (Plan)

1 Find the lateral area of each solid below.

6 points per question

（1） Triangular prism

8 in. 6 in. 5 in. 10 in.

⇒

8 in. 6 in. 5 in. 10 in.

(   )

（2） Rectangular prism

6 in. 8 in. 15 in.

⇒

(   )

（3） Cylinder

20 in. 5 in.

⇒

20 in. 5 in.

(   )

（4） Quadrilateral prism

12 in. 12 in. 10 in. 10 in.

⇒

12 in. 10 in. 10 in.

(   )

（5） Cone

12 in. 3 in.

⇒

12 in. 3 in.

(   )

（6） Cone

16 in. 4 in.

⇒

16 in. 4 in.

(   )

2 Find the surface area of each solid below.   8 points per question

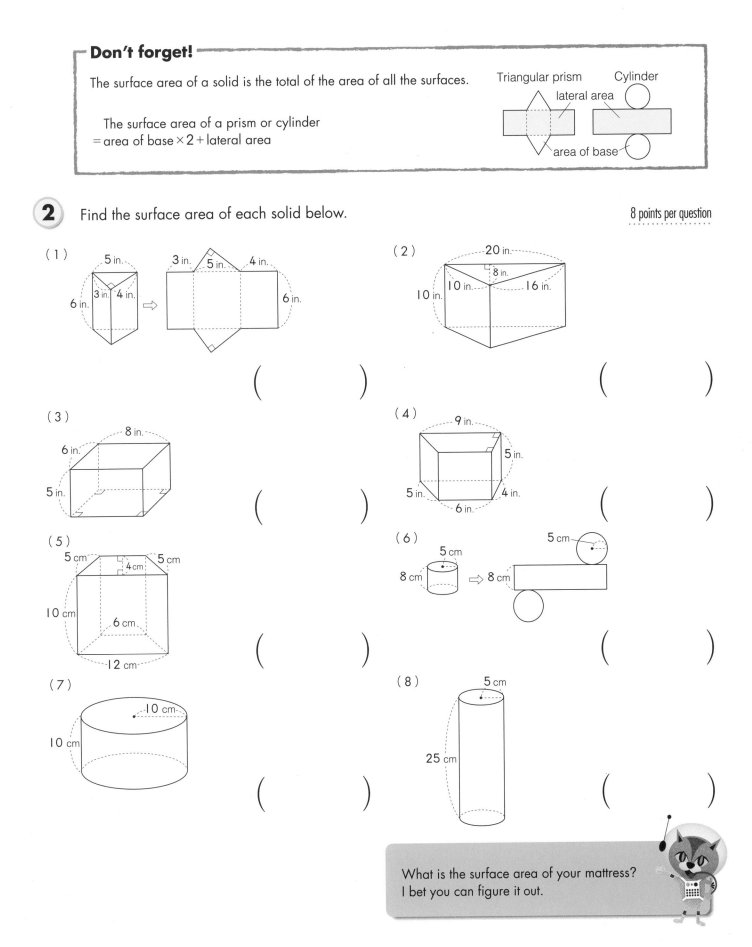

(1)
5 in.
6 in. 3 in. 4 in.
3 in. 5 in. 4 in.
6 in.

(   )

(2)
20 in.
8 in.
10 in. 10 in. 16 in.

(   )

(3)
8 in.
6 in.
5 in.

(   )

(4)
9 in.
5 in.
5 in. 4 in.
6 in.

(   )

(5)
5 cm 4 cm 5 cm
10 cm
6 cm
12 cm

(   )

(6)
5 cm
5 cm
8 cm   8 cm

(   )

(7)
10 cm
10 cm

(   )

(8)
5 cm
25 cm

(   )

What is the surface area of your mattress?
I bet you can figure it out.

© Kumon Publishing Co., Ltd.   53

27

Don't forget!

The surface area of a pyramid or cone = area of base + lateral area

Pyramid   Cone
lateral area

area of base

1 Find the surface area of each pyramid or cone below.

7 points per question

(1)

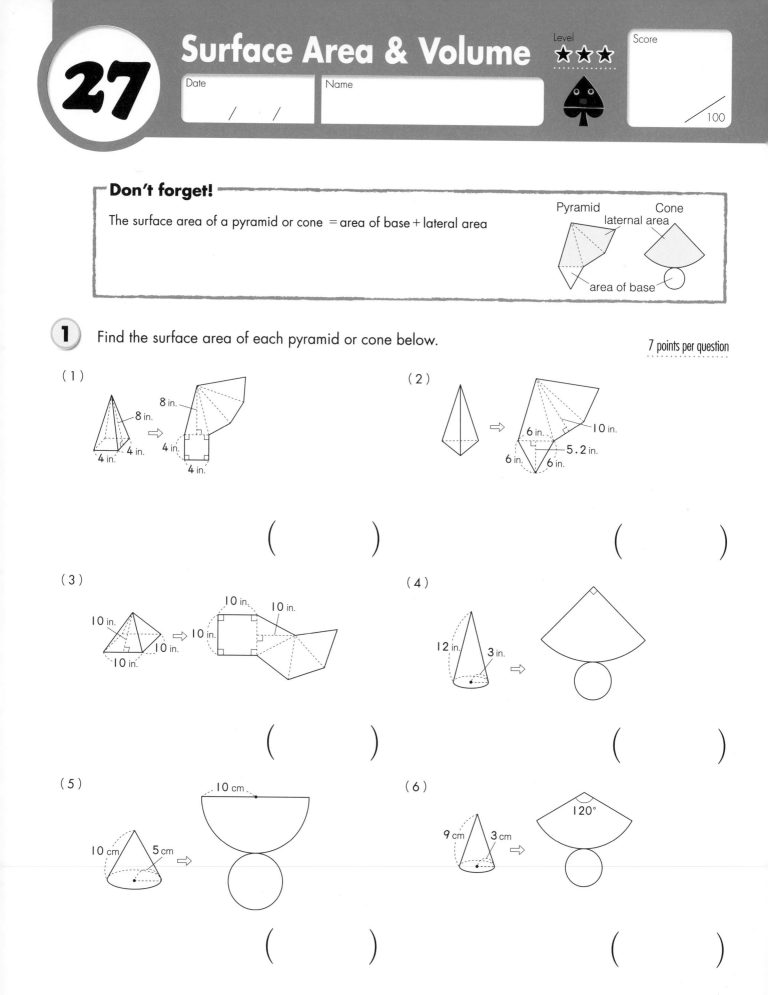

8 in.
8 in.
4 in.   4 in.
4 in.   4 in.
4 in.

( )

(2)

6 in.   10 in.
5.2 in.
6 in.   6 in.

( )

(3)

10 in.
10 in.   10 in.
10 in.
10 in.   10 in.

( )

(4)

12 in.   3 in.

( )

(5)

10 cm

10 cm   5 cm

( )

(6)

120°

9 cm   3 cm

( )

2 Answer the questions about the cylinder below.

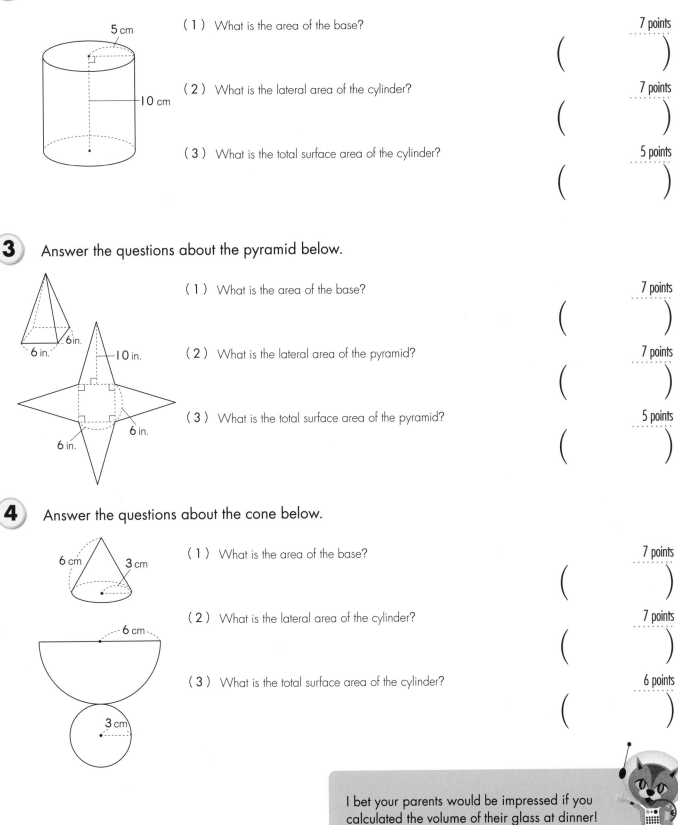

(1) What is the area of the base?   7 points

(   )

(2) What is the lateral area of the cylinder?   7 points

(   )

(3) What is the total surface area of the cylinder?   5 points

(   )

3 Answer the questions about the pyramid below.

(1) What is the area of the base?   7 points

(   )

(2) What is the lateral area of the pyramid?   7 points

(   )

(3) What is the total surface area of the pyramid?   5 points

(   )

4 Answer the questions about the cone below.

(1) What is the area of the base?   7 points

(   )

(2) What is the lateral area of the cylinder?   7 points

(   )

(3) What is the total surface area of the cylinder?   6 points

(   )

I bet your parents would be impressed if you calculated the volume of their glass at dinner!

1 The figures B-G are all drawn based on figure A. Answer the questions below.

10 points per question

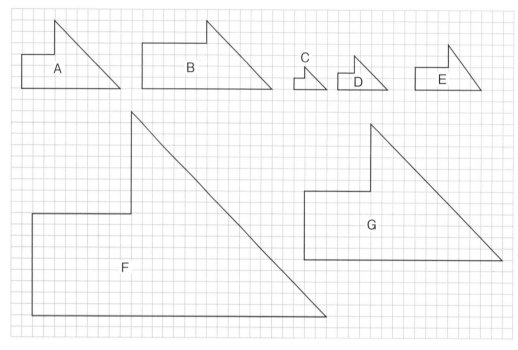

(1) Which figure has angles that all correspond to the angles in **A**, and sides that are twice the length of the sides of **A**?

( )

(2) Which figure has angles that all correspond to the angles in **A**, and sides that are **3** times the length of the sides of **A**?

( )

(3) Which figure has angles that all correspond to the angles in **A**, and sides that are half the length of the sides of **A**?

( )

(4) Which figure has angles that all correspond to the angles in **A**, and sides that are one third the length of the sides of **A**?

( )

Don't forget!

The sides are all twice as long.

☐ → ☐

A figure that has the same corresponding angles as the original shape, and all corresponding sides are twice as long is called a **twice-enlarged drawing**.

The sides are all half as long.

☐ → ☐

A figure that has the same corresponding angles as the original shape, and all the corresponding angles are half as long is called a **half-reduced drawing**.

2 Trapezoids B-F are all drawn based on figure A below. All the corresponding angles are equal.

15 points per question

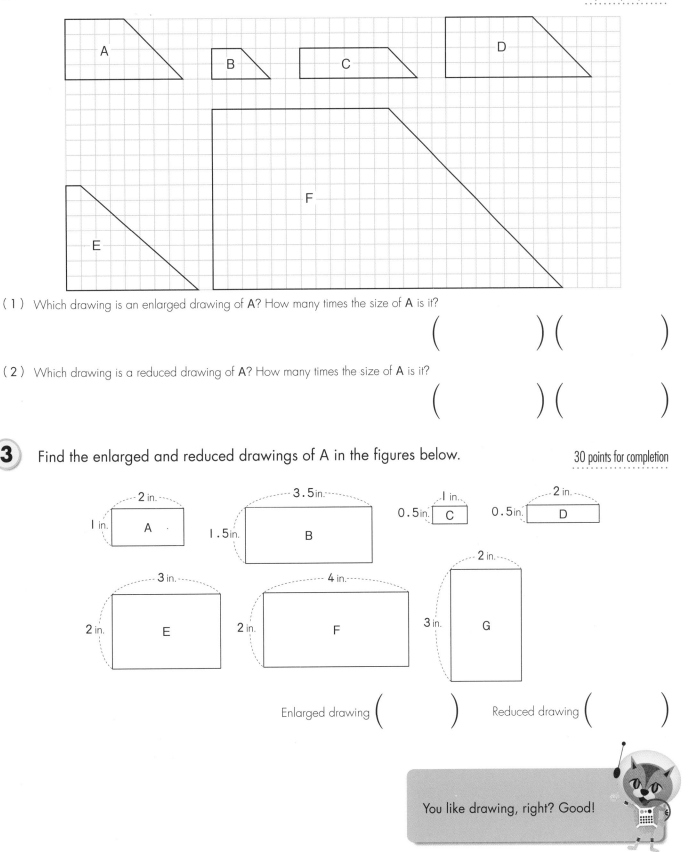

(1) Which drawing is an enlarged drawing of A? How many times the size of A is it?

(   ) (   )

(2) Which drawing is a reduced drawing of A? How many times the size of A is it?

(   ) (   )

3 Find the enlarged and reduced drawings of A in the figures below.

30 points for completion

Enlarged drawing (   )   Reduced drawing (   )

You like drawing, right? Good!

1 Triangle DEF is an enlarged drawing of triangle ABC. Answer the questions about the two triangles below.

4 points per question

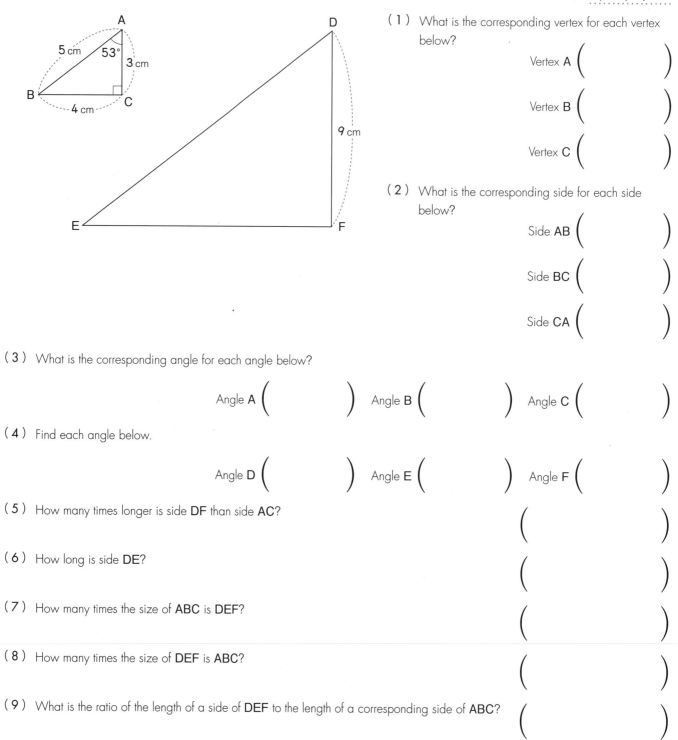

(1) What is the corresponding vertex for each vertex below?

Vertex A ( )

Vertex B ( )

Vertex C ( )

(2) What is the corresponding side for each side below?

Side AB ( )

Side BC ( )

Side CA ( )

(3) What is the corresponding angle for each angle below?

Angle A ( ) Angle B ( ) Angle C ( )

(4) Find each angle below.

Angle D ( ) Angle E ( ) Angle F ( )

(5) How many times longer is side DF than side AC?

( )

(6) How long is side DE?

( )

(7) How many times the size of ABC is DEF?

( )

(8) How many times the size of DEF is ABC?

( )

(9) What is the ratio of the length of a side of DEF to the length of a corresponding side of ABC?

( )

2 Figure **b** is an enlarged drawing of figure **a**. Answer the questions about the two shapes below.

8 points per question

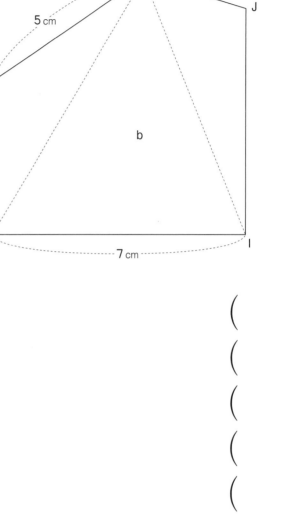

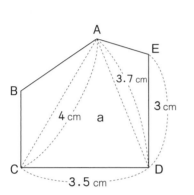

(1) How many times the size of **a** is **b**? (   )

(2) How many times the size of **b** is **a**? (   )

(3) How long is **AB**? (   )

(4) How long is **BC**? (   )

(5) How long is **IJ**? (   )

(6) What is the ratio of the length of a side in figure **a** to the length of a corresponding side in figure **b**? (   )

(7) The ratio between the lengths of the diagonals of the two shapes is equal to the ratio of corresponding sides of the two shapes. If the length of the diagonal **AC** is 4 centimeters, how long is the diagonal **FH**? (   )

(8) What is the length of the diagonal **FI**? (   )

You're doing a great job. Keep it up!

Scale Drawing

1 Draw a twice-enlarged drawing of the quadrilateral on the left.

10 points

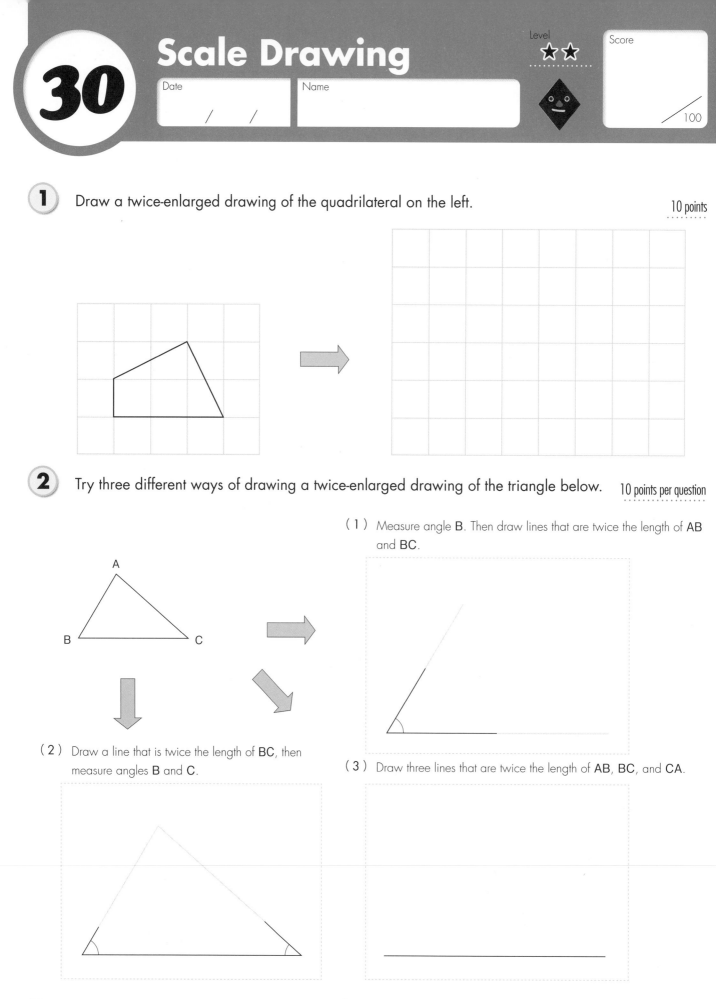

2 Try three different ways of drawing a twice-enlarged drawing of the triangle below. 10 points per question

(1) Measure angle **B**. Then draw lines that are twice the length of **AB** and **BC**.

(2) Draw a line that is twice the length of **BC**, then measure angles **B** and **C**.

(3) Draw three lines that are twice the length of **AB**, **BC**, and **CA**.

3 Try three different ways of drawing a half-reduced drawing of the triangle below.

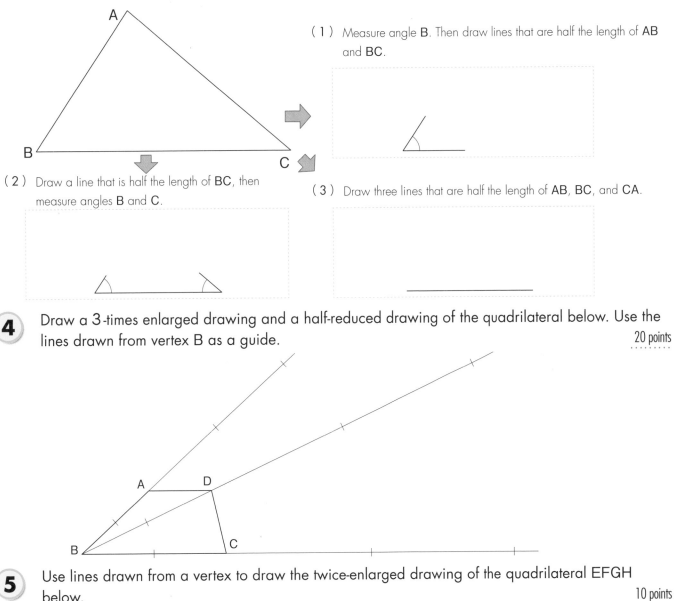

(1) Measure angle **B**. Then draw lines that are half the length of **AB** and **BC**.

(2) Draw a line that is half the length of **BC**, then measure angles **B** and **C**.

(3) Draw three lines that are half the length of **AB**, **BC**, and **CA**.

4 Draw a 3-times enlarged drawing and a half-reduced drawing of the quadrilateral below. Use the lines drawn from vertex B as a guide. 20 points

5 Use lines drawn from a vertex to draw the twice-enlarged drawing of the quadrilateral EFGH below. 10 points

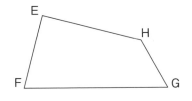

Architects do this kind of drawing very often!

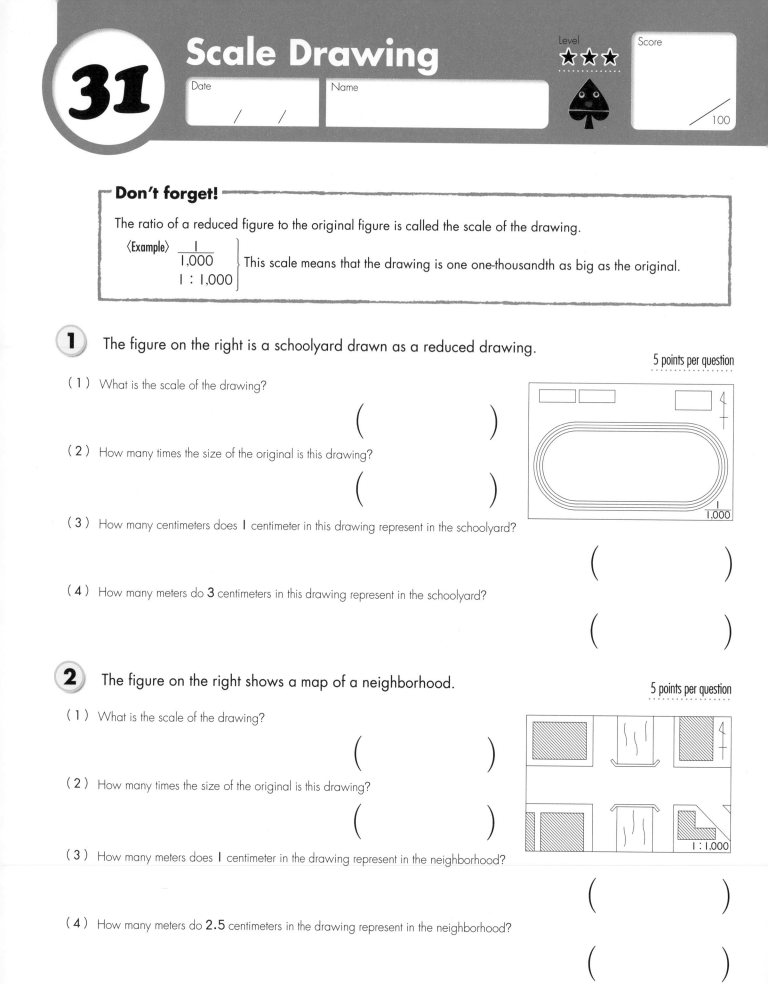

31

Scale Drawing

Date   /   /

Name

Level ★★★

Score   /100

Don't forget!

The ratio of a reduced figure to the original figure is called the scale of the drawing.

⟨Example⟩ $\frac{1}{1,000}$
1 : 1,000 } This scale means that the drawing is one one-thousandth as big as the original.

---

1 The figure on the right is a schoolyard drawn as a reduced drawing.

5 points per question

(1) What is the scale of the drawing?

(   )

(2) How many times the size of the original is this drawing?

(   )

(3) How many centimeters does 1 centimeter in this drawing represent in the schoolyard?

(   )

(4) How many meters do 3 centimeters in this drawing represent in the schoolyard?

(   )

$\frac{1}{1,000}$

2 The figure on the right shows a map of a neighborhood.

5 points per question

(1) What is the scale of the drawing?

(   )

(2) How many times the size of the original is this drawing?

(   )

(3) How many meters does 1 centimeter in the drawing represent in the neighborhood?

(   )

(4) How many meters do 2.5 centimeters in the drawing represent in the neighborhood?

(   )

1 : 1,000

62   © Kumon Publishing Co., Ltd.

3 How many meters in real life are represented by the length in each situation below?

4 points per question

(1) What is the original length of **2** centimeters in a ten-thousandths reduced drawing?

(   )

(2) What is the original length of **1.5** centimeters in a five-thousandths reduced drawing?

(   )

(3) What is the original length of **3** centimeters in a three-thousandths reduced drawing?

(   )

(4) What is the original length of **2** centimeters in a drawing with a scale of 1 : 10,000?

(   )

(5) What is the original length of **2.5** centimeters in a drawing with a scale of 1 : 5,000?

(   )

4 Write the scale of each map below as a fraction and a ratio.

10 points per question

(1) A road that is **20** meters long is drawn as 1 centimeter long on this map.

(,   )

(2) A road that is **500** meters long is drawn as 1 centimeter long on this map.

(,   )

(3) A road that is **1.2** kilometers long is drawn as **24** centimeters long on this map.

(,   )

(4) A road that is **3** kilometers long is drawn as **12** centimeters long on this map.

(,   )

Now you can read a map better! Excellent.

Date / /

Name

1 Answer the questions below by using figures A and B as scale models.

5 points per question

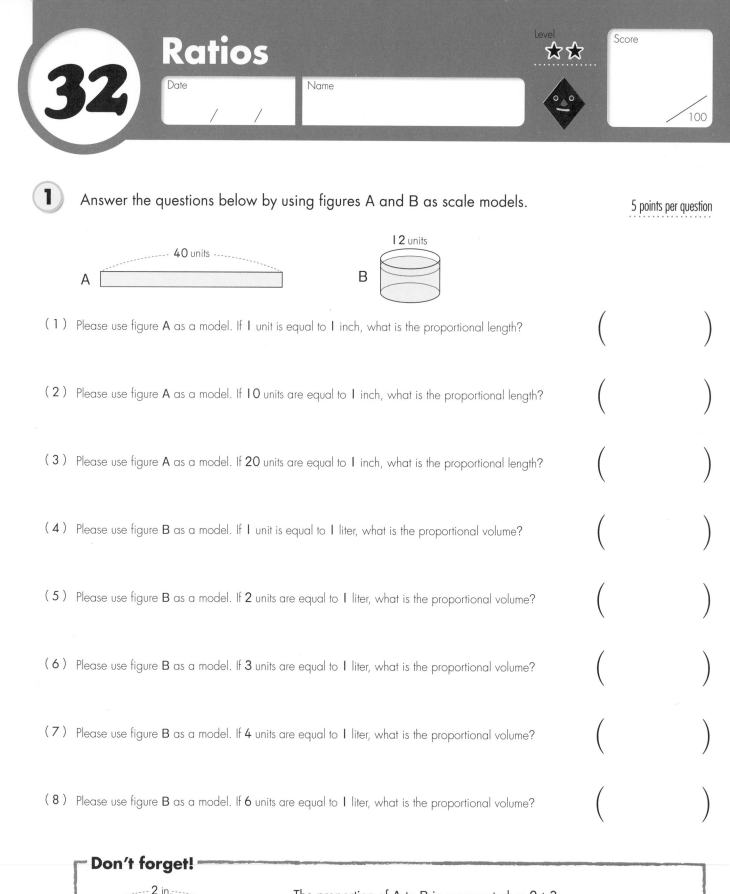

(1) Please use figure **A** as a model. If **I** unit is equal to **I** inch, what is the proportional length?

(   )

(2) Please use figure **A** as a model. If **10** units are equal to **I** inch, what is the proportional length?

(   )

(3) Please use figure **A** as a model. If **20** units are equal to **I** inch, what is the proportional length?

(   )

(4) Please use figure **B** as a model. If **I** unit is equal to **I** liter, what is the proportional volume?

(   )

(5) Please use figure **B** as a model. If **2** units are equal to **I** liter, what is the proportional volume?

(   )

(6) Please use figure **B** as a model. If **3** units are equal to **I** liter, what is the proportional volume?

(   )

(7) Please use figure **B** as a model. If **4** units are equal to **I** liter, what is the proportional volume?

(   )

(8) Please use figure **B** as a model. If **6** units are equal to **I** liter, what is the proportional volume?

(   )

Don't forget!

The proportion of A to B is represented as 2 : 3.
This is also called the ratio of the length of A to the length of B.

2 Write the ratio of each pair of measurements below.

8 points per question

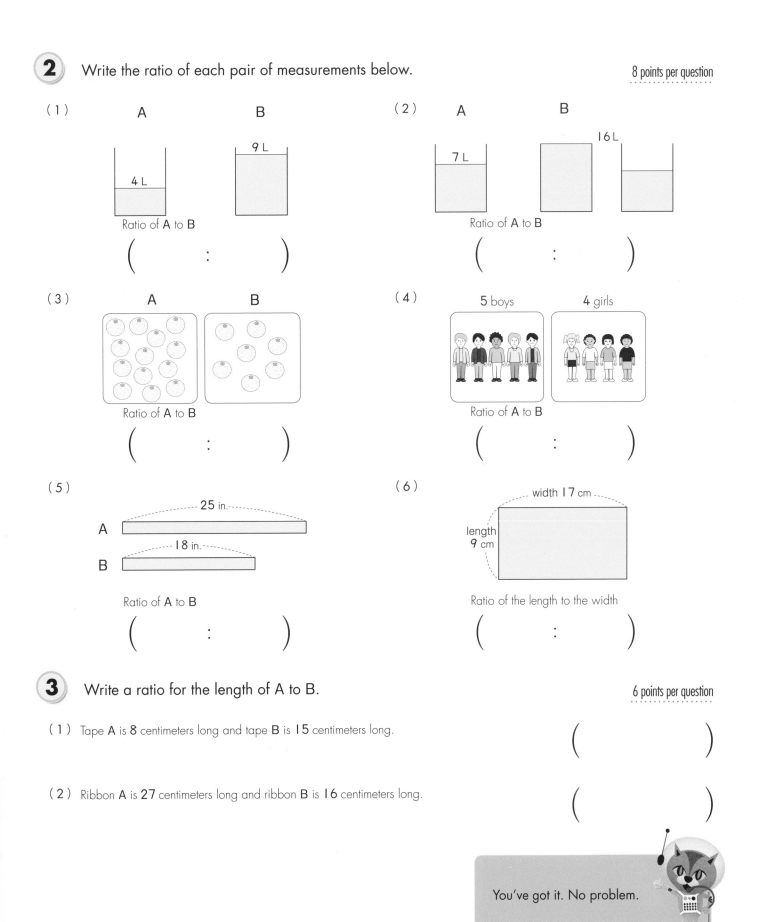

(1)　　A　　　　B

Ratio of A to B

(　　:　　)

(2)　　A　　　B

Ratio of A to B

(　　:　　)

(3)　　A　　　B

Ratio of A to B

(　　:　　)

(4)　　5 boys　　4 girls

Ratio of A to B

(　　:　　)

(5)

Ratio of A to B

(　　:　　)

(6)

Ratio of the length to the width

(　　:　　)

3 Write a ratio for the length of A to B.

6 points per question

(1) Tape A is 8 centimeters long and tape B is 15 centimeters long.

(　　　　)

(2) Ribbon A is 27 centimeters long and ribbon B is 16 centimeters long.

(　　　　)

You've got it. No problem.

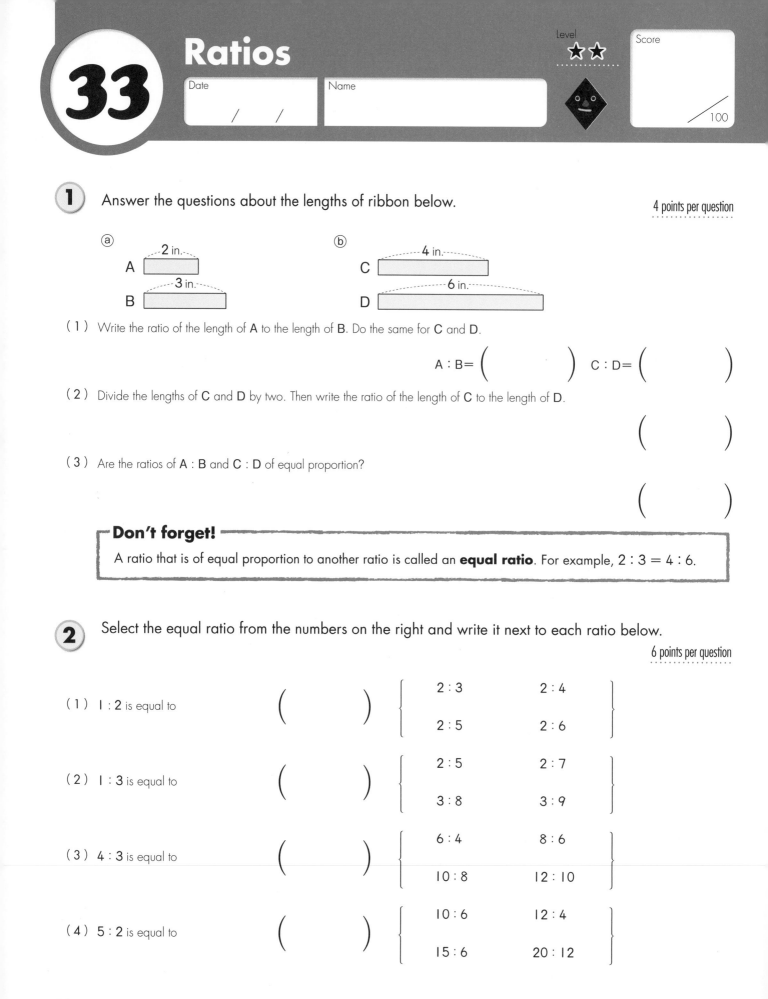

33 Ratios

Level ★★

Date / /

Name

Score /100

1 Answer the questions about the lengths of ribbon below.

4 points per question

ⓐ
2 in.
A

3 in.
B

ⓑ
4 in.
C

6 in.
D

(1) Write the ratio of the length of A to the length of B. Do the same for C and D.

A : B = ( ) C : D = ( )

(2) Divide the lengths of C and D by two. Then write the ratio of the length of C to the length of D.

( )

(3) Are the ratios of A : B and C : D of equal proportion?

( )

Don't forget!

A ratio that is of equal proportion to another ratio is called an **equal ratio**. For example, 2 : 3 = 4 : 6.

2 Select the equal ratio from the numbers on the right and write it next to each ratio below.

6 points per question

(1) 1 : 2 is equal to ( )
2 : 3   2 : 4
2 : 5   2 : 6

(2) 1 : 3 is equal to ( )
2 : 5   2 : 7
3 : 8   3 : 9

(3) 4 : 3 is equal to ( )
6 : 4   8 : 6
10 : 8   12 : 10

(4) 5 : 2 is equal to ( )
10 : 6   12 : 4
15 : 6   20 : 12

Don't forget!

$$\overset{\times 2}{2 : 3} = 4 : 6 \qquad \overset{\div 2}{4 : 6} = 2 : 3$$

$$\underset{\times 2}{} \qquad \underset{\div 2}{}$$

3 Write the appropriate number in each box below so that each pair of ratios is equal.

4 points per question

(1) $2 : 3 = 4 :$ ☐ (×2)

(2) $1 : 3 = 4 :$ ☐

(3) $2 : 3 = 6 :$ ☐

(4) $3 : 4 = 9 :$ ☐

(5) $4 : 3 =$ ☐ $: 12$

(6) $5 : 3 =$ ☐ $: 15$

(7) $3 : 7 =$ ☐ $: 21$

(8) $6 : 5 =$ ☐ $: 30$

(9) $3 : 9 = 1 :$ ☐ (÷3)

(10) $9 : 12 = 3 :$ ☐

(11) $8 : 12 = 2 :$ ☐

(12) $16 : 12 = 4 :$ ☐

(13) $6 : 8 =$ ☐ $: 4$

(14) $18 : 12 =$ ☐ $: 2$

(15) $27 : 18 =$ ☐ $: 2$

(16) $35 : 15 =$ ☐ $: 3$

Just a little further now. You can do it!

67

34

Ratios

1 In each case below, write how much of A you would have if B was 1.

5 points per question

（1）
A ⌐20 in.⌐
B ⌐20 in.⌐

(   )

（2）
A ⌐20 in.⌐
B ⌐10 in.⌐

(   )

（3）
A ⌐20 cm⌐
B ⌐40 cm⌐

(   )

（4）
A ⌐20 cm⌐
B ⌐30 cm⌐

(   )

Don't forget!

A ⌐20 in.⌐
B ⌐30 in.⌐

The **value of the ratio** is the first term divided by the second term.

$$A : B = 2 : 3 \quad 2 \div 3 = \frac{2}{3}$$

(When the value of ratio A : B is $\frac{2}{3}$, that means that A is $\frac{2}{3}$ if B is 1.)

---

2 Write the values of the ratios below.

3 points per question

（1） 1 : 3   (   )

（2） 5 : 7   (   )

（3） 3 : 8   (   )

（4） 4 : 6   (   )

（5） 9 : 12   (   )

（6） 3 : 1   (   )

（7） 7 : 5   (   )

（8） 8 : 3   (   )

（9） 24 : 36   (   )

（10） 28 : 35   (   )

Don't forget!

If the value of two ratios is equal, then the ratios are equal.

$2 : 3 \rightarrow$ the value of the ratio is $\dfrac{2}{3}$

$4 : 6 \rightarrow$ the value of the ratio is $\dfrac{2}{3}$

$\rightarrow 2 : 3 = 4 : 6$

$2 : 3 = 4 : 6$

| ratio of the length to the width | ratio of the length to the width |
| 2:3 | 4:6 |

3 Select the ratio that is equal to each ratio on the left.   5 points per question

(1) 1 : 3   (   )   { 2 : 4   2 : 5
                                2 : 6   2 : 7 }

(2) 3 : 5   (   )   { 4 : 6   6 : 8
                                8 : 10   9 : 15 }

4 Reduce the ratios below to simplest form.   4 points per question

$\div 2$

(1) 4 : 6 = [   ] : [   ]   (2) 4 : 8 = [:   ]

$\div 2$

(3) 12 : 16 = [:   ]   (4) 12 : 9 = [:   ]

(5) 24 : 42 = [:   ]   (6) 45 : 40 = [:   ]

5 Rewrite the values of ratios below as ratios in simplest form.   4 points per question

(1) $\dfrac{3}{5}$ (   )   (2) $1\dfrac{3}{11}$ (   )

(3) 0.5   (   )   (4) 0.25   (   )

Phew! You've done really well. Let's review!

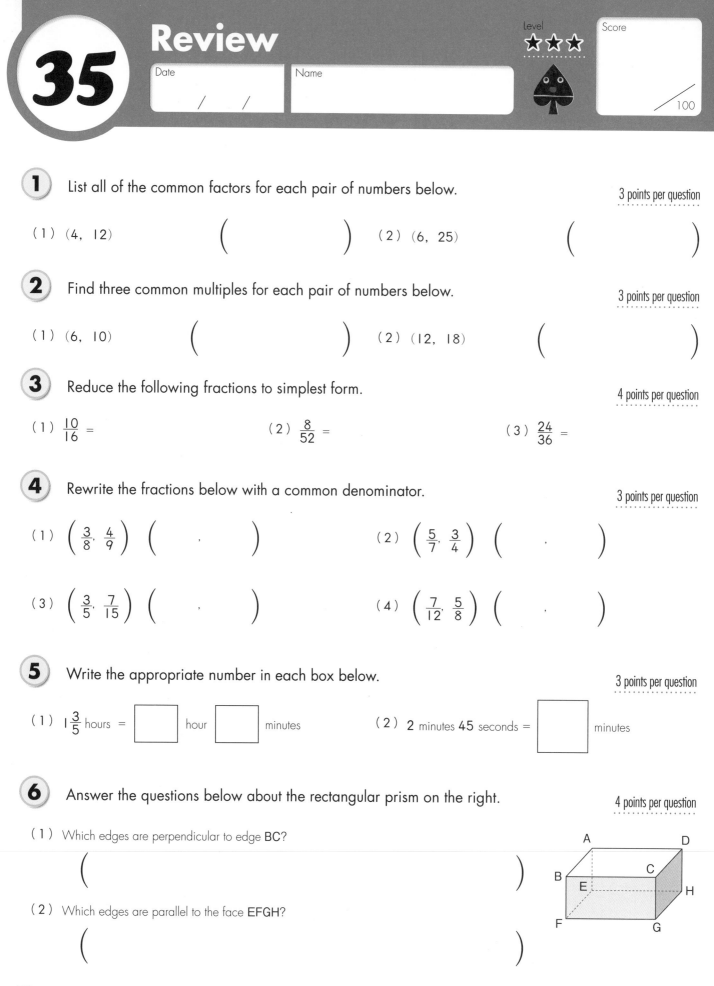

35 Review

Date / /

Name

Level ★★★

Score / 100

1 List all of the common factors for each pair of numbers below.

3 points per question

(1) (4, 12) ( ) (2) (6, 25) ( )

2 Find three common multiples for each pair of numbers below.

3 points per question

(1) (6, 10) ( ) (2) (12, 18) ( )

3 Reduce the following fractions to simplest form.

4 points per question

(1) $\frac{10}{16} =$ (2) $\frac{8}{52} =$ (3) $\frac{24}{36} =$

4 Rewrite the fractions below with a common denominator.

3 points per question

(1) $\left(\frac{3}{8}, \frac{4}{9} \right)$ (,) (2) $\left(\frac{5}{7}, \frac{3}{4} \right)$ (,)

(3) $\left(\frac{3}{5}, \frac{7}{15} \right)$ (,) (4) $\left(\frac{7}{12}, \frac{5}{8} \right)$ (,)

5 Write the appropriate number in each box below.

3 points per question

(1) $1\frac{3}{5}$ hours = ☐ hour ☐ minutes   (2) 2 minutes 45 seconds = ☐ minutes

6 Answer the questions below about the rectangular prism on the right.

4 points per question

(1) Which edges are perpendicular to edge **BC**?

( )

(2) Which edges are parallel to the face **EFGH**?

( )

7 Find the volume of each solid pictured below.

10 points per question

(1)

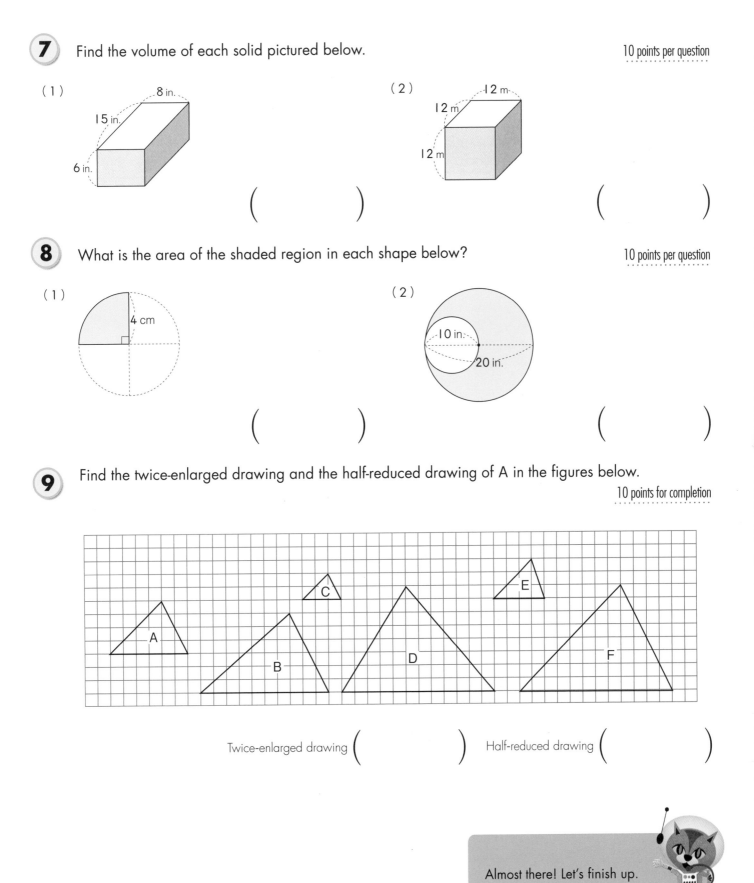

8 in.

15 in.

6 in.

(   )

(2)

12 m

12 m

12 m

(   )

8 What is the area of the shaded region in each shape below?

10 points per question

(1)

4 cm

(   )

(2)

10 in.

20 in.

(   )

9 Find the twice-enlarged drawing and the half-reduced drawing of A in the figures below.

10 points for completion

A

B

C

D

E

F

Twice-enlarged drawing (   ) Half-reduced drawing (   )

Almost there! Let's finish up.

36 Review

Level ★★★

Score

Date / /

Name

/100

1 Circle all of the multiples of 4.

5 points

16, 18, 22, 28, 34, 36, 46, 48

2 Find the GCF and the LCM for each pair of numbers below.

5 points per question

(1) (12, 18)

$\left(\text{GCF} \qquad , \text{LCM} \qquad\right)$

(2) (30, 50)

$\left(\text{GCF} \qquad , \text{LCM} \qquad\right)$

(3) (16, 24)

$\left(\text{GCF} \qquad , \text{LCM} \qquad\right)$

3 Reduce each fraction below to simplest form.

5 points per question

(1) $\frac{6}{9} =$

(2) $\frac{15}{25} =$

(3) $\frac{40}{48} =$

4 After rewriting each pair of fractions below with their common denominator, compare the numbers and write the larger fraction in the brackets.

5 points per question

(1) $\frac{3}{5}, \frac{4}{7}$

$\left(\qquad \right)$

(2) $\frac{5}{6}, \frac{7}{9}$

$\left(\qquad \right)$

5 The figure on the left is the solid built from the plan on the right.

5 points per question

(1) How long is edge IJ?

$\left(\qquad \right)$

3cm 2cm 5cm

A N
B M L K J
C F G H I
D E

(2) When you fold up the plan, which edge touches edge DE?

$\left(\qquad \right)$

(3) Which vertex touches vertex L?

$\left(\qquad \right)$

6 Find the volume of the solid on the right.

10 points

(   )

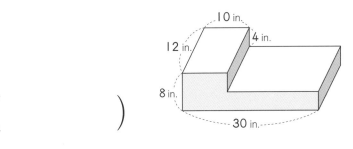

7 Write the appropriate number in each box below so that the ratios are equal.

2 points per question

(1) 2 : 3 = 6 : ☐

(2) 5 : 2 = 10 : ☐

(3) 4 : 7 = ☐ : 28

(4) 3 : 9 = 1 : ☐

(5) 8 : 12 = 2 : ☐

(6) 24 : 18 = ☐ : 3

8 Calculate the surface area and the volume of each figure below.

9 points per question

(1)

10 in.

20 in.

(2)

10 cm

4 cm

5 cm   8 cm

10 cm

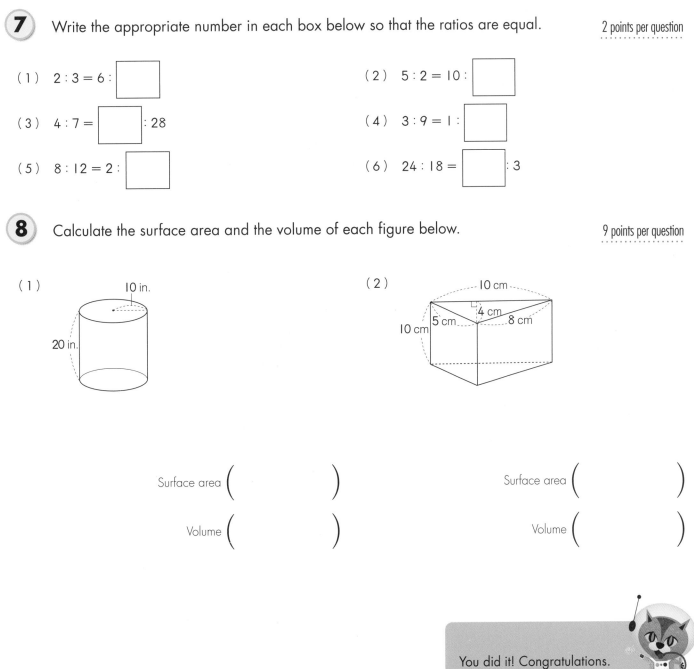

Surface area (   )

Volume (   )

Surface area (   )

Volume (   )

You did it! Congratulations.

1 Review
pp 2, 3

1 (Odd) 9, 11, 25, 41, 63, 87
(Even) 6, 18, 36, 50, 74

2 (1) 5.293 (2) 0.427

3 (1) $\frac{6}{7}$, $\frac{5}{7}$, $\frac{4}{7}$, $\frac{2}{7}$ (2) $\frac{3}{4}$, $\frac{3}{5}$, $\frac{3}{8}$, $\frac{3}{10}$

4 (1) $\frac{4}{7}$ (2) $\frac{6}{5}$ (3) $\frac{6}{9}$ (4) $\frac{8}{6}$

5 (1) 65°
(2) $180 - 65 = 115$ **Ans.** 115°

6 (1) 75°   (2) 7 cm

7 (1) $180 - 70 - 35 = 75$ **Ans.** 75°
(2) $360 - 90 - 80 - 75 = 115$ **Ans.** 115°

8 (1) $6 \times 7 = 42$ **Ans.** 42 in.²
(2) $16 \times 10 \div 2 = 80$ **Ans.** 80 in.²

2 Review
pp 4, 5

1 (1) 132   (2) 321

2 (1) 34   (2) 67   (3) 4.5   (4) 0.829

3 (1) 0.8   (2) 0.375   (3) $\frac{6}{10}$ (4) $\frac{108}{100}$

4 (1) $\frac{3}{5}$ (2) 1.27

5 (1) $(180 - 36) \div 2 = 72$ **Ans.** 72°
(2) $180 - 85 - 55 = 40$, $180 - 40 = 140$ **Ans.** 140°

6 (1) A, E   (2) A, B

7 $12 \times 6 \div 2 + 12 \times 6 \div 2 = 84$ **Ans.** 84 cm²

8 A

9 (1) 3   (2) 4   (3) 2   (4) 2

3 Least Common Multiple
pp 6, 7

1 3, 6, 9, 12, 15, 18

2 (1) 2, 4, 6, 8, 10   (2) 4, 8, 12, 16, 20
(3) 5, 10, 15, 20, 25   (4) 6, 12, 18, 24, 30

3 7, 14, 21, 28

4 8, 16, 24

5 9, 18, 27

6 18, 21, 24, 27

7 15, 48, 36, 51, 12, 66

8 24, 32, 36, 40, 44, 48, 52

9 35, 40, 45, 50, 55, 60

10 10

11 16

12 33

13 12

4 Least Common Multiple
pp 8, 9

1 16, 20, 24, 28, 32, 36, 40

2 18, 24, 30, 36, 42, 48, 54

3 18, 27, 36, 45, 54

4 12, 24, 36, 48, 60

5 18, 36, 54, 72

6 8, 16, 24, 32

7 (1) 18   (2) 8

8 (1) 24   (2) 12   (3) 6   (4) 12

9 (1) (6), (6), $\boxed{18}$ (2) (18), (18), $\boxed{18}$
(3) (9), (9), $\boxed{18}$

10 (1) 24   (2) 60   (3) 36   (4) 24

5 Greatest Common Factor
pp 10, 11

1 (1) 1, 2   (2) 1, 3   (3) 1, 2, 4   (4) 1, 5
(5) 1, 2, 3, 6 (6) 1, 7 (7) 1, 2, 4, 8
(8) 1, 3, 9 (9) 1, 2, 5, 10 (10) 1, 2, 3, 4, 6, 12

2 (1) 6   (2) 6   (3) 6

3 (1) 5   (2) 6   (3) 10

4 (1) 6   (2) 10

5 (1) 4   (2) 6   (3) 6   (4) 8   (5) 10   (6) 8

6 LCM & GCF
pp 12, 13

1 (1) 6   (2) 15   (3) 12   (4) 9   (5) 24   (6) 18
(7) 36   (8) 45   (9) 72   (10) 80

2 (1) 12   (2) 60   (3) 18   (4) 24   (5) 30   (6) 24

3 (1) 2   (2) 4   (3) 3   (4) 6   (5) 3   (6) 7
(7) 9   (8) 5   (9) 6   (10) 6   (11) 8   (12) 7
(13) 5   (14) 2   (15) 10   (16) 12   (17) 15   (18) 20
(19) 18   (20) 60

1 (1) ~ (4)

1	②	③	4	⑤	6	⑦	8	9	10
11	12	13	14	15	16	17	18	19	20
21	22	23	24	25	26	27	28	29	30
31	32	33	34	35	36	37	38	39	40

(5) 2, 3, 5, 7, 11, 13, 17, 19, 23, 29, 31, 37

2 (1) 1, 41 (2) Yes, 41 is prime number
(3) 1, 2, 3, 6, 7, 14, 21, 42
(4) No, 42 is not prime number.

3 43, 47, 59, 61, 67

4 (1) 3 (2) 5 (3) 2 (4) 3 (5) 3 (6) 2

5 (1) 2×5 (2) $2 \times 2 \times 3$ (3) $2 \times 3 \times 3$
(4) $2 \times 2 \times 5$ (5) 3×7 (6) 5×5
(7) $3 \times 3 \times 3$ (8) $2 \times 2 \times 7$ (9) $3 \times 3 \times 5$
(10) $2 \times 2 \times 13$ (11) $2 \times 2 \times 3 \times 5$ (12) $2 \times 3 \times 11$

8 Fractions

1 (1) 1 (2) 2 (3) 6 (4) 1 (5) 2 (6) 9 (7) 1
(8) 2 (9) 12

2 (1) 2 (2) 2 (3) 1 (4) 1 (5) 1

3 (1) 1 (2) 2 (3) 3 (4) 1 (5) 2 (6) 3 (7) 1
(8) 2 (9) 3 (10) 1 (11) 2 (12) 3 (13) 1 (14) 2
(15) 3 (16) 2 (17) 3 (18) 2 (19) 3 (20) 5

9 Fractions

1 (1) $\frac{1}{3}$ (2) $\frac{1}{4}$ (3) $\frac{1}{6}$ (4) $\frac{1}{3}$ (5) $\frac{1}{4}$ (6) $\frac{3}{4}$ (7) $\frac{1}{2}$
(8) $\frac{1}{4}$ (9) $\frac{3}{5}$ (10) $\frac{1}{2}$ (11) $\frac{1}{3}$ (12) $\frac{4}{5}$ (13) $\frac{1}{2}$ (14) $\frac{1}{3}$
(15) $\frac{4}{5}$ (16) $\frac{1}{2}$ (17) $\frac{2}{3}$ (18) $\frac{4}{5}$ (19) $\frac{1}{2}$ (20) $\frac{1}{3}$ (21) $\frac{4}{5}$
(22) $\frac{1}{2}$ (23) $\frac{3}{4}$ (24) $\frac{2}{5}$ (25) $\frac{5}{4}$ (26) $\frac{4}{3}$ (27) $\frac{5}{3}$ (28) $\frac{3}{2}$
(29) $\frac{1}{4}$ (30) $\frac{1}{3}$

2 (1) 2 (2) 3 (3) 4 (4) 2 (5) 4 (6) 6
(7) 2 (8) 6 (9) 9 (10) 2 (11) 6 (12) 16
(13) 2 (14) 15 (15) 2 (16) 6 (17) 32 (18) 10
(19) 12 (20) 44

10 Fractions

1 (1) $\left(\frac{3}{12},\ \frac{8}{12}\right)$ (2) $\left(\frac{4}{12},\ \frac{9}{12}\right)$ (3) $\left(\frac{5}{20},\ \frac{8}{20}\right)$
(4) $\left(\frac{12}{30},\ \frac{5}{30}\right)$ (5) $\left(\frac{9}{12},\ \frac{10}{12}\right)$ (6) $\left(\frac{3}{18},\ \frac{4}{18}\right)$
(7) $\left(\frac{9}{24},\ \frac{10}{24}\right)$ (8) $\left(\frac{9}{15},\ \frac{7}{15}\right)$ (9) $\left(\frac{9}{6},\ \frac{8}{6}\right)$
(10) $\left(\frac{28}{20},\ \frac{25}{20}\right)$ (11) $\left(\frac{20}{12},\ \frac{19}{12}\right)$ (12) $\left(\frac{28}{24},\ \frac{27}{24}\right)$

2 (1) $\frac{2}{3}\left(\frac{1}{2}=\frac{3}{6},\ \frac{2}{3}=\frac{4}{6}\right)$ (2) $\frac{3}{4}\left(\frac{2}{3}=\frac{8}{12},\ \frac{3}{4}=\frac{9}{12}\right)$
(3) $\frac{4}{5}\left(\frac{3}{4}=\frac{15}{20},\ \frac{4}{5}=\frac{16}{20}\right)$ (4) $\frac{7}{8}\left(\frac{6}{7}=\frac{48}{56},\ \frac{7}{8}=\frac{49}{56}\right)$
(5) $\frac{3}{5}\left(\frac{3}{5}=\frac{6}{10},\ \frac{1}{2}=\frac{5}{10}\right)$ (6) $\frac{4}{9}\left(\frac{1}{3}=\frac{3}{9},\ \frac{4}{9}\right)$
(7) $\frac{3}{4}\left(\frac{7}{12},\ \frac{3}{4}=\frac{9}{12}\right)$ (8) $\frac{11}{15}\left(\frac{3}{5}=\frac{9}{15},\ \frac{11}{15}\right)$
(9) $\frac{5}{8}\left(\frac{5}{8}=\frac{15}{24},\ \frac{7}{12}=\frac{14}{24}\right)$ (10) $\frac{8}{9}\left(\frac{8}{9}=\frac{16}{18},\ \frac{5}{6}=\frac{15}{18}\right)$
(11) $\frac{4}{3}\left(\frac{3}{4}=\frac{9}{12},\ \frac{4}{3}=\frac{16}{12}\right)$ (12) $\frac{5}{4}\left(\frac{6}{5}=\frac{24}{20},\ \frac{5}{4}=\frac{25}{20}\right)$
(13) $\frac{5}{3}\left(\frac{3}{2}=\frac{9}{6},\ \frac{5}{3}=\frac{10}{6}\right)$ (14) $\frac{11}{6}\left(\frac{11}{6}=\frac{55}{30},\ \frac{9}{5}=\frac{54}{30}\right)$
(15) $\frac{9}{8}\left(\frac{9}{8}=\frac{18}{16},\ \frac{17}{16}\right)$ (16) $\frac{5}{4}\left(\frac{5}{4}=\frac{15}{12},\ \frac{7}{6}=\frac{14}{12}\right)$

11 Fractions

1 (1) 1 (2) $\frac{1}{2}$ (3) $\frac{1}{4}$ (4) $\frac{1}{3}$ (5) $\frac{1}{6}$ (6) $\frac{1}{12}$ (7) $\frac{1}{60}$
(8) $\frac{3}{4}$ (9) $1\frac{5}{6}$ (10) $3\frac{5}{12}$

2 (1) 60 (2) 1 (3) 17 (4) 2 (5) 3 (6) 4 (7) 6
(8) 10 (9) 50 (10) 12 (11) 48 (12) 15 (13) 45 (14) 20
(15) 8, 40

3 (1) 1 (2) $\frac{1}{2}$ (3) $\frac{1}{4}$ (4) $\frac{1}{3}$ (5) $\frac{1}{6}$ (6) $\frac{1}{12}$ (7) $\frac{1}{60}$
(8) $\frac{3}{4}$ (9) $1\frac{1}{5}$ (10) $5\frac{2}{15}$

4 (1) 60 (2) 1 (3) 59 (4) 2 (5) 3 (6) 4 (7) 28
(8) 6 (9) 10 (10) 12 (11) 48 (12) 15 (13) 45 (14) 20
(15) 4, 40

12 Circles

1 (1) $3 \times 3.14 = 9.42$ — **Ans.** 9.42 in.
(2) $5 \times 3.14 = 15.7$ — **Ans.** 15.7 in.
(3) $10 \times 3.14 = 31.4$ — **Ans.** 31.4 in.
(4) $2 \times 2 = 4$, $4 \times 3.14 = 12.56$ — **Ans.** 12.56 cm
(5) $4 \times 2 = 8$, $8 \times 3.14 = 25.12$ — **Ans.** 25.12 cm
(6) $4.5 \times 2 = 9$, $9 \times 3.14 = 28.26$ — **Ans.** 28.26 cm

2 (1) $6 \times 3.14 = 18.84$ — **Ans.** 18.84 in.
(2) $7 \times 2 \times 3.14 = 43.96$ — **Ans.** 43.96 in.
(3) $4.2 \times 3.14 = 13.188$ — **Ans.** 13.188 in.
(4) $10 \times 3.14 \div 2 + 10 = 25.7$ — **Ans.** 25.7 in.
(5) $10 \times 3.14 + 10 = 41.4$ — **Ans.** 41.4 cm
(6) $10 \times 3.14 + 20 \times 2 = 71.4$ — **Ans.** 71.4 cm
(7) $10 \times 3.14 \div 2 + 10 \times 3 = 45.7$ — **Ans.** 45.7 cm
(8) $20 \times 3.14 + 20 \times 2 = 102.8$ — **Ans.** 102.8 cm

13 Circles
pp 26, 27

1
(1) $1 \times 1 \times 3.14 = 3.14$ **Ans.** 3.14 in.²
(2) $3 \times 3 \times 3.14 = 28.26$ **Ans.** 28.26 in.²
(3) $5 \times 5 \times 3.14 = 78.5$ **Ans.** 78.5 in.²
(4) $4 \div 2 = 2$, $2 \times 2 \times 3.14 = 12.56$ **Ans.** 12.56 cm²
(5) $8 \div 2 = 4$, $4 \times 4 \times 3.14 = 50.24$ **Ans.** 50.24 cm²
(6) $12 \div 2 = 6$, $6 \times 6 \times 3.14 = 113.04$ **Ans.** 113.04 cm²

2
(1) $3 \times 3 \times 3.14 \div 2 = 14.13$ **Ans.** 14.13 in.²
(2) $4 \div 2 = 2$, $2 \times 2 \times 3.14 \div 2 = 6.28$ **Ans.** 6.28 in.²
(3) $8 \div 2 = 4$, $4 \times 4 \times 3.14 \div 2 = 25.12$ **Ans.** 25.12 cm²
(4) $10 \div 2 = 5$, $5 \times 5 \times 3.14 \div 2 = 39.25$ **Ans.** 39.25 cm²
(5) $2 \times 2 \times 3.14 \div 4 = 3.14$ **Ans.** 3.14 in.²
(6) $3 \times 3 \times 3.14 \div 4 = 7.065$ **Ans.** 7.065 in.²
(7) $6 \times 6 \times 3.14 \div 4 = 28.26$ **Ans.** 28.26 cm²
(8) $5 \times 5 \times 3.14 \div 4 = 19.625$ **Ans.** 19.625 cm²

14 Circles
pp 28, 29

1
(1) $6 \times 6 \times 3.14 - 4 \times 4 \times 3.14 = 62.8$ **Ans.** 62.8 in.²
(2) $4 \div 2 = 2$, $4 \times 4 \times 3.14 - 2 \times 2 \times 3.14 = 37.68$
   Ans. 37.68 in.²
(3) $12 \div 2 = 6$, $10 \div 2 = 5$, $2 \div 2 = 1$,
$6 \times 6 \times 3.14 - 5 \times 5 \times 3.14 - 1 \times 1 \times 3.14 = 31.4$
   Ans. 31.4 in.²
(4) $8 \div 2 = 4$, $4 \times 4 \times 3.14 + 8 \times 10 = 130.24$
   Ans. 130.24 cm²
(5) $10 \div 2 = 5$, $5 \times 5 \times 3.14 \div 2 \times 4 + 10 \times 10 = 257$
   Ans. 257 cm²

2
(1) $15 \div 2 = 7.5$, $15 \times 15 - 7.5 \times 7.5 \times 3.14 = 48.375$
   Ans. 48.375 in.²
(2) $20 \div 2 = 10$, $20 \times 20 - 10 \times 10 \times 3.14 = 86$
   Ans. 86 in.²
(3) $20 \div 2 = 10$, $10 \div 2 = 5$,
$10 \times 10 \times 3.14 \div 2 - 5 \times 5 \times 3.14 = 78.5$
   Ans. 78.5 in.²
(4) $10 \div 2 = 5$, $8 \div 2 = 4$,
$5 \times 5 \times 3.14 \div 2 - 4 \times 4 \times 3.14 \div 2 = 14.13$
   Ans. 14.13 cm²
(5) $14 \div 2 = 7$,
$14 \times 14 \times 3.14 \div 4 - 7 \times 7 \times 3.14 \div 2 = 76.93$
   Ans. 76.93 cm²
(6) $16 \div 2 = 8$, $12 \div 2 = 6$, $4 \div 2 = 2$,
$8 \times 8 \times 3.14 \div 2 + 6 \times 6 \times 3.14 \div 2 + 2 \times 2 \times 3.14 \div 2$
$= 163.28$ **Ans.** 163.28 cm²

15 Circles & Sectors
pp 30, 31

1   A, D, F, G

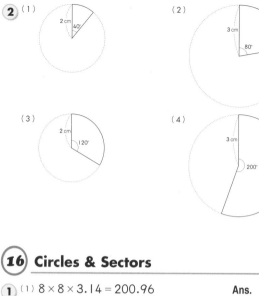

2 (1)   (2)   (3)   (4)

16 Circles & Sectors
pp 32, 33

1
(1) $8 \times 8 \times 3.14 = 200.96$ **Ans.** 200.96 cm²
(2) $45 \div 360 = \frac{1}{8}$ **Ans.** $\frac{1}{8}$
(3) $200.96 \times \frac{1}{8} = 25.12$ **Ans.** 25.12 cm²

2
(1) $3 \times 3 \times 3.14 \times \frac{60}{360} = 4.71$ **Ans.** 4.71 in.²
(2) $9 \times 9 \times 3.14 \times \frac{150}{360} = 105.975$ **Ans.** 105.975 in.²

3
(1) $\frac{1}{4}$
(2) $2 \times 3 \times 3.14 \times \frac{1}{4} = 4.71$ **Ans.** 4.71 in.
(3) $\frac{1}{6}$
(4) $\frac{1}{3}$
(5) $3 \times 2 + 4.71 = 10.71$ **Ans.** 10.71 in.
(6) $3 \times 2 + 2 \times 3 \times 3.14 \times \frac{1}{6} = 9.14$ **Ans.** 9.14 in.

17 Rectangular Prisms
pp 34, 35

1   (Rectangular prism) B, D   (Cube) F

2

	Rectangular prism	Cube
Number of faces	6	6
Number of edges	12	12
Number of vertices	8	8

3
(1) edge DC,   edge EF,   edge HG
(2) edge BC,   edge FG,   edge EH
(3) edge BF,   edge CG,   edge DH
(4) 2 faces   (5) 2 faces   (6) 2 faces

④ (1) edge DC, edge HG, edge EF, edge BF, edge CG,
edge DH, edge AE, edge AD, edge BC, edge FG,
edge EH

(2) 6 faces

⑱ Rectangular Prisms
pp 36,37

① (1) edge AD, edge AE
(2) edge BA, edge BC, edge EF, edge FG
(3) edge DC, edge HG, edge EF
(4) edge AD, edge EH, edge FG

② (1) edge AE, edge BF, edge CG, edge DH
(2) edge AB, edge DC, edge EF, edge HG

③ (1) face BFGC, face CGHD, face DHEA
(2) face ABCD, face BFGC, face FGHE, face DHEA
(3) face ABCD, face AEFB, face EFGH, face CGHD

④ (1) face DCGH   (2) face AEHD

⑤ (1) edge CG, edge DH, edge CD, edge GH
(2) edge AE, edge DH, edge AD, edge EH

⑲ Rectangular Prisms
pp 38,39

① (1) B   (2) A

② (1) B, C   (2) A, E

③ (1) 12 cm   (2) 5 cm   (3) 8 cm   (4) edge I H
(5) edge DC   (6) edge GF   (7) vertex E   (8) vertex N
(9) vertex I, vertex K   (10) face EDGF
(11) face ANCB, face MJGD, face KJML, face EDGF

⑳ Rectangular Prisms
pp 40,41

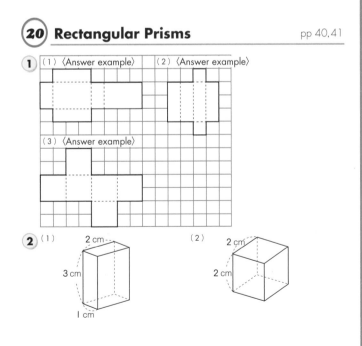

① (1) 〈Answer example〉   (2) 〈Answer example〉

(3) 〈Answer example〉

② (1)   (2)

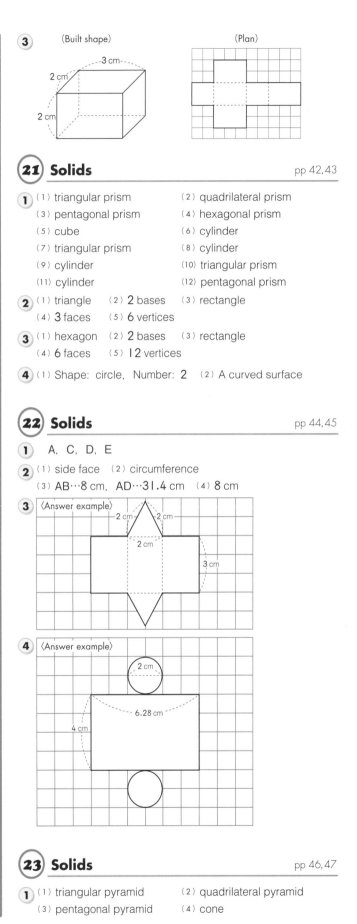

③ (Built shape)   (Plan)
3 cm
2 cm
2 cm

㉑ Solids
pp 42,43

① (1) triangular prism   (2) quadrilateral prism
(3) pentagonal prism   (4) hexagonal prism
(5) cube   (6) cylinder
(7) triangular prism   (8) cylinder
(9) cylinder   (10) triangular prism
(11) cylinder   (12) pentagonal prism

② (1) triangle   (2) 2 bases   (3) rectangle
(4) 3 faces   (5) 6 vertices

③ (1) hexagon   (2) 2 bases   (3) rectangle
(4) 6 faces   (5) 12 vertices

④ (1) Shape: circle, Number: 2   (2) A curved surface

㉒ Solids
pp 44,45

① A, C, D, E

② (1) side face   (2) circumference
(3) AB···8 cm, AD···31.4 cm   (4) 8 cm

③ 〈Answer example〉
2 cm — 2 cm
2 cm
3 cm

④ 〈Answer example〉
2 cm
6.28 cm
4 cm

㉓ Solids
pp 46,47

① (1) triangular pyramid   (2) quadrilateral pyramid
(3) pentagonal pyramid   (4) cone

(2) (1) square (2) I (3) isosceles triangle (4) 4
(3) (1) Shape: circle, Number: I (2) A curved surface
(4) A, B, D
(5) A, C

(24) Surface Area & Volume pp 48, 49

(1)
(1) $8 \times 5 \div 2 = 20$ **Ans.** 20 in.^{2}
(2) $9 \times 2 \div 2 = 9$ **Ans.** 9 in.^{2}
(3) $(4 + 6) \times 5 \div 2 = 25$ **Ans.** 25 in.^{2}
(4) $10 \times 10 \times 3.14 = 314$ **Ans.** 314 in.^{2}
(5) $5 \times 5 \times 3.14 = 78.5$ **Ans.** 78.5 in.^{2}
(6) $10 \times 8.6 \div 2 = 43$ **Ans.** 43 in.^{2}
(7) $10 \times 10 = 100$ **Ans.** 100 cm^2
(8) $8 \times 8 \times 3.14 = 200.96$ **Ans.** 200.96 cm^2
(9) $4 \times 4 \times 3.14 = 50.24$ **Ans.** 50.24 cm^2

(2)
(1) $4 \times 5 \times 8 = 160$ **Ans.** 160 in.^{3}
(2) $10 \times 10 \times 10 = 1,000$ **Ans.** 1,000 in.^{3}
(3) $8 \times 12 \div 2 \times 10 = 480$ **Ans.** 480 in.^{3}
(4) $10 \times 3 \div 2 \times 8 = 120$ **Ans.** 120 in.^{3}
(5) $10 \times 10 \times 3.14 \times 20 = 6,280$ **Ans.** 6,280 cm^3
(6) $5 \times 5 \times 3.14 \times 8 = 628$ **Ans.** 628 cm^3
(7) $(12 + 18) \times 10 \div 2 \times 12 = 1,800$ **Ans.** 1,800 cm^3
(8) $25 \times 8 = 200$ **Ans.** 200 cm^3

(25) Surface Area & Volume pp 50, 51

(1)
(1) $6 \times 6 \times 10 \times \frac{1}{3} = 120$ **Ans.** 120 in.^{3}
(2) $6 \times 6 \times 18 \times \frac{1}{3} = 216$ **Ans.** 216 in.^{3}
(3) $6 \times 5.2 \div 2 \times 10 \times \frac{1}{3} = 52$ **Ans.** 52 in.^{3}
(4) $16 \times 6 \times \frac{1}{3} = 32$ **Ans.** 32 cm^3
(5) $30 \times 9 \times \frac{1}{3} = 90$ **Ans.** 90 cm^3
(6) $25 \times 12 \times \frac{1}{3} = 100$ **Ans.** 100 cm^3

(2)
(1) $6 \times 6 \times 3.14 \times 8 \times \frac{1}{3} = 301.44$ **Ans.** 301.44 in.^{3}
(2) $3 \times 3 \times 3.14 \times 10 \times \frac{1}{3} = 94.2$ **Ans.** 94.2 in.^{3}
(3) $10 \times 10 \times 3.14 \times 15 \times \frac{1}{3} = 1,570$ **Ans.** 1,570 in.^{3}
(4) $9 \times 9 \times 3.14 \times 6 \times \frac{1}{3} = 508.68$ **Ans.** 508.68 in.^{3}
(5) $5 \times 5 \times 3.14 \times 9 \times \frac{1}{3} = 235.5$ **Ans.** 235.5 cm^3
(6) $4 \times 4 \times 3.14 \times 15 \times \frac{1}{3} = 251.2$ **Ans.** 251.2 cm^3
(7) $200 \times 18 \times \frac{1}{3} = 1,200$ **Ans.** 1,200 cm^3
(8) $18 \times 11 \times \frac{1}{3} = 66$ **Ans.** 66 cm^3

(26) Surface Area & Volume pp 52, 53

(1)
(1) $10 \times (6 + 8 + 5) = 190$ **Ans.** 190 in.^{2}
(2) $15 \times (6 \times 2 + 8 \times 2) = 420$ **Ans.** 420 in.^{2}
(3) $20 \times 5 \times 2 \times 3.14 = 628$ **Ans.** 628 in.^{2}
(4) $10 \times 12 \div 2 \times 4 = 240$ **Ans.** 240 in.^{2}
(5) $12 \times 12 \times 3.14 \div 4 = 113.04$ **Ans.** 113.04 in.^{2}
(6) $16 \times 16 \times 3.14 \div 4 = 200.96$ **Ans.** 200.96 in.^{2}

(2)
(1) $4 \times 3 \div 2 \times 2 = 12$, $6 \times (3 + 5 + 4) = 72$,
$12 + 72 = 84$ **Ans.** 84 in.^{2}
(2) $20 \times 8 \div 2 \times 2 = 160$, $10 \times (10 + 16 + 20) = 460$,
$160 + 460 = 620$ **Ans.** 620 in.^{2}
(3) $6 \times 8 \times 2 = 96$, $5 \times (6 \times 2 + 8 \times 2) = 140$,
$96 + 140 = 236$ **Ans.** 236 in.^{2}
(4) $(6 + 9) \times 4 \div 2 \times 2 = 60$, $5 \times (4 + 6 + 5 + 9) = 120$,
$60 + 120 = 180$ **Ans.** 180 in.^{2}
(5) $(6 + 12) \times 4 \div 2 \times 2 = 72$, $10 \times (6 + 5 + 12 + 5) = 280$,
$72 + 280 = 352$ **Ans.** 352 cm^2
(6) $5 \times 5 \times 3.14 \times 2 = 157$, $8 \times 5 \times 2 \times 3.14 = 251.2$,
$157 + 251.2 = 408.2$ **Ans.** 408.2 cm^2
(7) $10 \times 10 \times 3.14 \times 2 = 628$, $10 \times 10 \times 2 \times 3.14 = 628$,
$628 + 628 = 1,256$ **Ans.** 1,256 cm^2
(8) $5 \times 5 \times 3.14 \times 2 = 157$, $25 \times 5 \times 2 \times 3.14 = 785$,
$157 + 785 = 942$ **Ans.** 942 cm^2

(27) Surface Area & Volume pp 54, 55

(1)
(1) $4 \times 4 = 16$, $4 \times 8 \div 2 \times 4 = 66$,
$16 + 64 = 80$ **Ans.** 80 in.^{2}
(2) $6 \times 5.2 \div 2 = 15.6$, $6 \times 10 \div 2 \times 3 = 90$,
$15.6 + 90 = 105.6$ **Ans.** 105.6 in.^{2}
(3) $10 \times 10 = 100$, $10 \times 10 \div 2 \times 4 = 200$,
$100 + 200 = 300$ **Ans.** 300 in.^{2}
(4) $3 \times 3 \times 3.14 = 28.26$, $12 \times 12 \times 3.14 \div 4 = 113.04$,
$28.26 + 113.04 = 141.3$ **Ans.** 141.3 in.^{2}
(5) $5 \times 5 \times 3.14 = 78.5$, $10 \times 10 \times 3.14 \div 2 = 157$,
$78.5 + 157 = 235.5$ **Ans.** 235.5 cm^2
(6) $3 \times 3 \times 3.14 = 28.26$, $9 \times 9 \times 3.14 \div 3 = 84.78$,
$28.26 + 84.78 = 113.04$ **Ans.** 113.04 cm^2

(2)
(1) $5 \times 5 \times 3.14 = 78.5$ **Ans.** 78.5 cm^2
(2) $10 \times 5 \times 2 \times 3.14 = 314$ **Ans.** 314 cm^2
(3) $78.5 \times 2 + 314 = 471$ **Ans.** 471 cm^2

(3)
(1) $6 \times 6 = 36$ **Ans.** 36 in.^{2}
(2) $6 \times 10 \div 2 \times 4 = 120$ **Ans.** 120 in.^{2}
(3) $36 + 120 = 156$ **Ans.** 156 in.^{2}

(4)
(1) $3 \times 3 \times 3.14 = 28.26$ **Ans.** 28.26 cm^2
(2) $6 \times 6 \times 3.14 \div 2 = 56.52$ **Ans.** 56.52 cm^2
(3) $28.26 + 56.52 = 84.78$ **Ans.** 84.78 cm^2

28 Scale Drawing

pp 56,57

1 (1) G (2) F (3) D (4) C

2 (1) F, 3 times (2) B, $\frac{1}{2}$ times

3 (Enlarged drawing) F, (Reduced drawing) C

29 Scale Drawing

pp 58,59

1 (1) Vertex A→Vertex D, Vertex B→Vertex E,
Vertex C→Vertex F
(2) Side AB→Side DE, Side BC→Side EF,
Side CA→Side FD
(3) Angle A→Angle D, Angle B→Angle E,
Angle C→Angle F
(4) Angle D→53°, Angle E→37°, Angle F→90°
(5) 3 times (6) 15 cm (7) 3 times
(8) $\frac{1}{3}$ times (9) 3 : 1

2 (1) 2 times (double) (2) $\frac{1}{2}$ times (3) 2.5 cm
(4) 2 cm (5) 6 cm (6) 1 : 2 (7) 8 cm
(8) 7.4 cm

30 Scale Drawing

pp 60,61

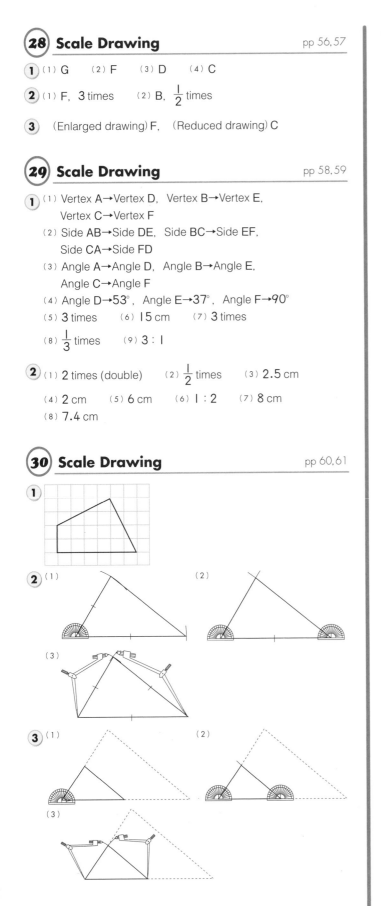

4

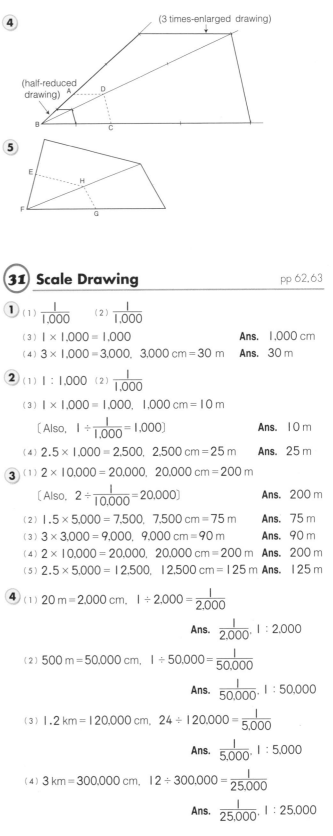

(3 times-enlarged drawing)

(half-reduced drawing)

5

31 Scale Drawing

pp 62,63

1 (1) $\frac{1}{1,000}$ (2) $\frac{1}{1,000}$
(3) $1 \times 1,000 = 1,000$ **Ans.** 1,000 cm
(4) $3 \times 1,000 = 3,000$, 3,000 cm = 30 m **Ans.** 30 m

2 (1) 1 : 1,000 (2) $\frac{1}{1,000}$
(3) $1 \times 1,000 = 1,000$, 1,000 cm = 10 m
[Also, $1 \div \frac{1}{1,000} = 1,000$] **Ans.** 10 m
(4) $2.5 \times 1,000 = 2,500$, 2,500 cm = 25 m **Ans.** 25 m

3 (1) $2 \times 10,000 = 20,000$, 20,000 cm = 200 m
[Also, $2 \div \frac{1}{10,000} = 20,000$] **Ans.** 200 m
(2) $1.5 \times 5,000 = 7,500$, 7,500 cm = 75 m **Ans.** 75 m
(3) $3 \times 3,000 = 9,000$, 9,000 cm = 90 m **Ans.** 90 m
(4) $2 \times 10,000 = 20,000$, 20,000 cm = 200 m **Ans.** 200 m
(5) $2.5 \times 5,000 = 12,500$, 12,500 cm = 125 m **Ans.** 125 m

4 (1) 20 m = 2,000 cm, $1 \div 2,000 = \frac{1}{2,000}$
Ans. $\frac{1}{2,000}$, 1 : 2,000
(2) 500 m = 50,000 cm, $1 \div 50,000 = \frac{1}{50,000}$
Ans. $\frac{1}{50,000}$, 1 : 50,000
(3) 1.2 km = 120,000 cm, $24 \div 120,000 = \frac{1}{5,000}$
Ans. $\frac{1}{5,000}$, 1 : 5,000
(4) 3 km = 300,000 cm, $12 \div 300,000 = \frac{1}{25,000}$
Ans. $\frac{1}{25,000}$, 1 : 25,000

32 Ratios

1 (1) 40 inches (2) 4 inches (3) 2 inches (4) 12 L
(5) 6 L (6) 4 L (7) 3 L (8) 2 L

2 (1) 4 : 9 (2) 7 : 16 (3) 13 : 7
(4) 5 : 4 (5) 25 : 18 (6) 9 : 17

3 (1) 8 : 15 (2) 27 : 16

33 Ratios
pp 66,67

1 (1) A : B = (2 : 3), C : D = (4 : 6)
(2) 2 : 3 (3) equal proportion

2 (1) 2 : 4 (2) 3 : 9 (3) 8 : 6 (4) 15 : 6

3 (1) 6 (2) 12 (3) 9 (4) 12
(5) 16 (6) 25 (7) 9 (8) 36
(9) 3 (10) 4 (11) 3 (12) 3
(13) 3 (14) 3 (15) 3 (16) 7

34 Ratios
pp 68,69

1 (1) 1 (2) 2 (3) $\frac{1}{2}$ (4) $\frac{2}{3}$

2 (1) $\frac{1}{3}$ (2) $\frac{5}{7}$ (3) $\frac{3}{8}$ [Also, 0.375] (4) $\frac{2}{3}$
(5) $\frac{3}{4}$ [Also, 0.75] (6) 3
(7) $\frac{7}{5}$ [Also, $1\frac{2}{5}$ or 1.4] (8) $\frac{8}{3}$ [Also, $2\frac{2}{3}$]
(9) $\frac{2}{3}$ (10) $\frac{4}{5}$ [Also, 0.8]

3 (1) 2 : 6 (2) 9 : 15

4 (1) 2 : 3 (2) 1 : 2 (3) 3 : 4 (4) 4 : 3
(5) 4 : 7 (6) 9 : 8

5 (1) 3 : 5 (2) 14 : 11 (3) 1 : 2 (4) 1 : 4

35 Review
pp 70,71

1 (1) 1, 2, 4 (2) 1

2 (1) 30, 60, 90 (2) 36, 72, 108

3 (1) $\frac{5}{8}$ (2) $\frac{2}{13}$ (3) $\frac{2}{3}$

4 (1) $\left(\frac{27}{72}, \frac{32}{72}\right)$ (2) $\left(\frac{20}{28}, \frac{21}{28}\right)$
(3) $\left(\frac{9}{15}, \frac{7}{15}\right)$ (4) $\left(\frac{14}{24}, \frac{15}{24}\right)$

5 (1) 1, 36 (2) $2\frac{3}{4}\left(\frac{11}{4}\right)$

6 (1) edge AB, edge DC, edge FB, edge GC
(2) edge AB, edge DC, edge BC, edge AD

7 (1) $15 \times 8 \times 6 = 720$ Ans. 720 in.³
(2) $12 \times 12 \times 12 = 1,728$ Ans. 1,728 cm³

8 (1) $4 \times 4 \times 3.14 \div 4 = 12.56$ Ans. 12.56 cm²
(2) $10 \times 10 \times 3.14 - 5 \times 5 \times 3.14 = 235.5$
  Ans. 235.5 in.²

9 Twice-enlarged drawing : F
Half-reduced drawing : C

36 Review
pp 72,73

1 16, 28, 36, 48

2 (1) (GCF) 6, (LCM) 36
(2) (GCF) 10, (LCM) 150
(3) (GCF) 8, (LCM) 48

3 (1) $\frac{2}{3}$ (2) $\frac{3}{5}$ (3) $\frac{5}{6}$

4 (1) $\frac{3}{5}\left(\frac{3}{5} = \frac{21}{35}, \frac{4}{7} = \frac{20}{35}\right)$
(2) $\frac{5}{6}\left(\frac{5}{6} = \frac{15}{18}, \frac{7}{9} = \frac{14}{18}\right)$

5 (1) 3 cm (2) edge HG (3) vertex N

6 $12 \times 10 \times 4 + 12 \times 30 \times (8 - 4) = 1,920$
  Ans. 1,920 in.³
[Also, $12 \times 30 \times 8 - 12 \times (30 - 10) \times 4 = 1,920$]

7 (1) 9 (2) 4 (3) 16 (4) 3 (5) 3 (6) 4

8 (1) (surface area)
  $10 \times 10 \times 3.14 \times 2 + 20 \times 10 \times 2 \times 3.14 = 1,884$
  Ans. 1,884 in.²
  (volume)
  $10 \times 10 \times 3.14 \times 20 = 6,280$
  Ans. 6,280 in.³
(2) (surface area)
  $10 \times 4 \div 2 \times 2 = 40$, $10 \times (5 + 8 + 10) = 230$
  $40 + 230 = 270$ Ans. 270 cm²
  (volume)
  $10 \times 4 \div 2 \times 10 = 200$ Ans. 200 cm³